The Gerbil in Behavioral Investigations

Del Thiessen is professor of psychology at the University of Texas at Austin. Pauline Yahr is assistant professor of psychobiology at the University of California at Irvine.

The Dan Danciger Publication Series

Courtesy D. Robinson.

The Gerbil in Behavioral Investigations

Mechanisms of Territoriality and Olfactory Communication

by

Del Thiessen

and

Pauline Yahr

University of Texas Press/Austin and London

Library of Congress Cataloging in
      Publication Data

Thiessen, Delbert D
The gerbil in behavioral investigations.

Bibliography: p.
Includes index.
1. Meriones unguiculatus—Behavior.
      2. Territoriality (Zoology) 3. Animal
      communication. 4. Smell.   I. Yahr,
      Pauline, 1946– joint author.   II. Title.
QL737.R638T48   599'.3233   76–13834
ISBN 0–292–72709–7

Contents

Tables

Figures

Acknowledgments

The research described in this monograph
was supported by the following grants: MH
14076, MH 8304, MH 25425, NSF-USDP
GU-1598 University of Texas Biomedical
Sciences Research Grant, and an award to
the Center for Behavior Genetics from the
Russell Sage Foundation. Wherever possible, we have acknowledged participation
in this research by credits in the text. Susan
Goar participated in the construction of the
brain atlas, and Ping Shen contributed in-
valuable service on many aspects of these
problems. Dr. Robert Sanders of the University of Texas at Austin Department of
Zoology collaborated on the immunological
research. We additionally wish to thank
Nancy Belknap and Janet Carter for typing
and editing much of the manuscript.

We wish to thank Robert Selander of the
Department of Zoology at the University of
Texas for doing the electrophoretic anal-
yses.

Introduction

Many species, especially birds and several mammalian forms, display territorial behavior. Territorial behavior may be essential in these species for the integrity of social relations as well as necessary for the uninterrupted flow of species-specific behaviors from generation to generation. Still, it must be recognized that territorial behavior is not a universal characteristic of life. Many other forms of behavior maintain adequate social relations and preserve the gene pool. These range from personal distance responses and linear hierarchies of social order to complex mixtures of social differentiation and responsibility. Territorial defense, as one aspect of territorial behavior, is widespread and often encompasses a diversity of habits found in other social systems. What is more, the defense of areas may be an advanced form of social behavior and a forerunner of even more complex systems. Therefore it deserves our attention.

The most notable progress in the understanding of territoriality has been made in the wild (Carpenter 1958; Calhoun 1963; Lack 1966; Howard 1920; Wynne-Edwards 1962). Data have been reported and significant theoretical advances have been made. However, little concerted effort has focused on the details of territoriality. The reason for this is that the majority of investigations have been conducted in the field, particularly with birds, where the systematic manipulation of variables is difficult or impossible. It has been said many times that ornithological observations are well ahead of experimental manipulations. As a result, a great many notions about how territories are formed, maintained, and used for survival have never been adequately tested. Concepts remain vague, anthropomorphic, and often highly speculative—leading to unusually wide and colorful generalizations based more on personal biases than concrete analyses (e.g., Ardrey 1966; Lorenz 1966).

In this monograph we do not interpret territoriality in the strict sense of active defense of specific areas. Rather, we broaden the meaning to include all social activities restricted to living areas. Any behavior that is more likely to take place in a circumscribed living area than elsewhere is as much a part of territorial behavior as active defense. Active defense may be a primary component of territoriality in many species, but it need not be the defining characteristic. In our opinion notions of territoriality must be enlarged to encompass a host of adaptive behaviors performed in localized areas.

Laboratory studies could substantiate or refute current ideas and add new theoretical support to concepts of territoriality. Regrettably, such studies have lagged far behind field investigations. This investigative lag is apparent for at least two reasons: (a) the meaning of territoriality under laboratory conditions has never been clearly defined, and (b) the domesticated species usually studied indoors do not lend themselves to naturalistic observations or interpretation (Beach 1950; Lockard 1968;

Lorenz 1950). What is needed, it seems, is a well-balanced integration of laboratory investigations coupled with the use of species that show the most potential for demonstrating facets of territorial behavior. For these reasons our laboratory group has turned to the Mongolian gerbil, *Meriones unguiculatus*, for study.

The Mongolian gerbil combines qualities necessary for laboratory work along with discrete territorial responses associated with a midventral scent gland and other forms of chemosignaling. It has been under domestic care for only a few years and shares behavioral and physiological characteristics with many species toward which we may wish to extrapolate—for example, higher primates. The combined qualities of ease of management, scent gland communication, and semidomestication make the Mongolian gerbil an ideal representative for social and chemosocial investigations.

The broad intention of this monograph is to lay a fundamental, factual, and theoretical base for the elucidation of territorial behavior in a single species. Specifically, we wish to (a) define the characteristics of the territorial response, (b) describe the environmental regulation of the behavior, (c) unravel the social (particularly sexual) significance of the behavior, and (d) uncover the physiological and genetic mechanisms of control. Our goal is to ask and answer the experimental questions necessary to establish the Mongolian gerbil as a prototype for the study of territoriality in mammalian species. When this is done, social characteristics of several species can be meaningfully compared and appropriate generalizations formed. Advances made in these directions during the past few years are summarized here.

The crux of understanding territoriality in the Mongolian gerbil lies in a thorough knowledge of the scent marking response in this and other species. Gerbils person-

alize their environment by pressing a midventral sebaceous gland on objects, leaving a sebum which is oily to the touch and musky in odor. They also secrete a Harderian pheromone from the nares and perhaps use urine deposition and other forms of olfactory communication. While the primary focus here is on ventral scent marking, other forms of olfactory communication will also be stressed. With the use of scent marking and olfactory signaling of other sorts, social messages are distributed along chemical gradients. In mammals, especially, chemosignals can be used for a variety of purposes. Pheromones often evoke stereotyped activities in insects. In contrast, mammals show more flexibility in their behavior and even a great deal of learning about the meaning of olfactory signals. Gerbil pheromones, for example, do not function exclusively in barrier defense but are also involved in other complex social events, such as reproductive activities. Territorial behavior and scent marking must be viewed in this larger context and not restricted to area defense.

Scent marking, the primary subject of this monograph, is intimately related to the sex, social status, and reproductive capacity of the individual. The more we can learn about the control of scent marking, the more we will discover about the function of chemical messages and the meaning of territoriality. It is essential to determine when, and under what conditions, the gerbil will mark; it is necessary to discover the precise signaling qualities of the deposited sebum; and it is imperative that we find out what physiological mechanisms regulate the behavior. Finally, we must deduce the evolutionary history of these relations and point out their adaptive value. While we can only approximate these goals here, it is hoped that this monograph will lead to a better understanding of the diversity of scent marking, pheromone communication, and territoriality.

Chapter 1 is of general importance for the understanding of territoriality and chemical communication. A brief review of current thought is included, with a discussion of species that communicate with olfactory signals (pheromones) within a territorial context and species whose behavior depends upon hormone control. The Mongolian gerbil is one of these species. A survey of the literature permits many inferences regarding the significance of territoriality. The most important may be that territoriality is a primary mechanism of evolution, not for those holding an area, but for those unable to achieve this luxury. Living in a territory is indeed a luxury, for it protects individuals and populations from the pull of natural selection. Exiled individuals, on the other hand, radiate out to less advantageous positions and are under harsh selection pressure. If the phenotypic characteristics which lead to the loss of a territory happen to match those required for survival in a new environment, then a new genotype-environment relation is established and evolution can proceed in a new direction.

In Chapter 2 the general characteristics of the Mongolian gerbil are specified, including its taxonomic position, physiological features and behavioral attributes. A survey of behavioral studies of the Mongolian gerbil is presented. The scent gland complexes found in this and other species are discussed within an endocrinological, morphological, and behavioral framework. This chapter is meant to lay the foundation for a clearer understanding of later discussions, as well as to acquaint the reader with what we believe is a superb research animal. There is little question that the Mongolian gerbil will someday become one of the principal laboratory species of many behaviorists.

Chapter 3 discusses our findings on the social significance of territorial behavior and scent marking in the Mongolian gerbil. Chapter 4 emphasizes the hormone mechanisms that contribute to the behavior, and Chapter 5 details what we know about the central regulation of territorial marking and includes a complete stereotaxic atlas of the adult gerbil brain. Chapter 6 reviews recent findings on Harderian secretion, suggesting that this pheromonal system is of great importance in territorial function. Finally, Chapter 7 outlines an overall model of territorial behavior, which integrates features of chemoreception, gene induction, hormone intervention, brain activity, social interaction, and evolution. The model is most applicable to the Mongolian gerbil with its scent communication; however, it is general enough to have relevance to other species which are equally territorial, but which may use different communicative devices to establish their environmental claims. The model should suggest relevant comparative investigations.

The Gerbil in Behavioral Investigations

1

Scent Marking, Chemocommunication, and Gland Characteristics

Cutaneous Scent Glands and Social Communication

The nearly universal use of chemical communication is suggested by the frequent appearance of scent glands and pheromone secretions in many species. According to Müller-Schwarze (1967) scent glands have been described in fifteen of the nineteen mammalian orders, and on the basis of location on the body, as many as forty different types can be classified. One species may possess a varied assortment of glands. Lagomorphs, for example, have chin glands, anal glands, paired inguinal glands, and Harderian, infraorbital and lachrymal glands situated in the orbit (Mykytowycz 1965; 1966a; 1966c; 1970). Arctic and Columbian ground squirrels have scent glands at the corner of the mouth, on the dorsal surface, and in the anal area (Steiner 1974). The flying phalanger possesses a major frontal gland and smaller glandular areas near the ears and at the angle of the eye (Schultze-Westrum 1965). The Mongolian gerbil has a ventral scent gland pad and diffusely organized sebaceous tissue on the neck. All vertebrate species except fish and marine organisms have Harderian glands located at the back of the eyeball.

Secretory cells, constituting the scent glands used in chemical communication, are of two types: holocrine, as in the Mongolian gerbil, and apocrine, as in the European rabbit. Holocrine glands form sebum from the total breakdown of secretory cells. On the other hand, apocrine glands synthesize sebum without cellular disintegration and pass the sebum out of the cell. Subcutaneous hair canals often serve as pathways for sebum exit into the environment for both types.

Apocrine and holocrine glands may appear in separate species suspected of sharing similar behaviors. For instance, apocrine glands appear in the chin complex of the rabbit (*Oryctolagus cuniculus*) and hare (Leporidae) and the cheeks of the pika (*Ochotona princeps*), whereas holocrine glands appear on the midventral surface and under the chin of the Mongolian gerbil (*Meriones unguiculatus*) and on the flank of the hamster (*Mesocricetus auratus*). On the other hand, the two general types may occur together, as in the inguinal glands of the rabbit and hare or the lateral gland of the short-tailed shrew (*Blarina brevicauda*). In the rabbit (*Oryctolagus cuniculus*) the apocrine chin glands secrete large amounts of proteins and carbohydrates (Goodrich and Mykytowycz 1972). These are relatively nonvolatile and could act as a short-distance olfactory or gustatory cue. The anal gland and portions of the inguinal glands have high amounts of lipid material, suitable as airborne signals. Presumably such variations in secretory materials could vary the chemical message. While not supported by strong evidence, Schaffer (1940) suggests that apocrine glands convey species-specific odors, whereas holocrine glands affect sexual behaviors. According to this speculation, a mixture of the two gland types would be most evident in reproductive processes related to individual recognition and courting behavior.

Most species have not been sufficiently studied to conclude that the presence of scent glands automatically indicates that they have behavioral significance. In fact, the early study of scent glands was concerned with the commercial use of civet, musk, and castoreum in perfumes. Civet is obtained primarily from Abyssinian species of civet cats; musk is obtained from a dorsal gland of the musk deer (*Moschus moschiferus*) in the Himalayas of southern China; and castoreum is obtained from the anal glands of the Siberian beaver (*Castor fiber*) and the Canadian beaver (*Castor canadensis*). The commercial value of these olfactory stimulants led to early advances in the investigation of scent glands (Moncrieff 1951) and the rearing of relevant animals.

Table 1 lists those species that possess discrete scent glands and associated behaviors—behaviors presumably of social significance. Eight orders involving 56 species have these characteristics. Certainly, many more species will emerge that use scent glands for chemical communication; and, even among those listed in Table 1, only the European rabbit, Columbian ground squirrel, sugar glider, black-tailed deer, Mongolian gerbil, and golden hamster have been investigated to any extent (see Thiessen and Rice 1976 for a review). For the most part, the significance of scent glands and associated behaviors remains cryptic.

Many additional means of chemically marking the environment have evolved and are indicated in Table 2. Anal rubbing or dragging is quite frequent, as are urination (including urine washing) and defecation. Less frequently used methods of marking include salivation and cloacal, cheek, and vaginal rubbing. Once these species have been studied more thoroughly, it should not be surprising to find that most of the marking traits have social significance.

It is notable that many of the products used for marking are byproducts of normal metabolic activities. Apparently, exudations often take on signaling capacities because they are present in large amounts as metabolic by-products and can assume secondary functions as the result of natural selection. The argument will be made later that scent communication and thermoenergetic functions have evolved as an integrated system to satisfy energetic requirements. The relationships are evident in prokaryotic as well as eukaryotic organisms. Similarly, reproductive variations result in a host of biochemical changes that can be selected for signaling capacities. For example, the body chemistry of the female changes radically from the follicular to the luteal phase of the estrous cycle, leading to metabolic by-products ideal for chemosignals. Likewise, the onset of puberty in both sexes offers a wide array of chemical changes that could act as social signals.

Ultimately, several criteria must be met before it can be definitively said that a scent gland has behavioral, evolutionary, and social significance. In ascending order of significance, it would seem that the following criteria must be established:

1.

The gland should appear well organized and interface with the substratum that the animal usually contacts. Thus, glands on sides, flanks, and dorsal surface occur in a number of small rodents that either inhabit burrows or use runways through thick vegetation (Ewer 1968). Among species that commonly sandbathe (e.g., *Heteromyid*) or depress their anal-genital area against the substrate because of any irritation (e.g., Carnivora and many rodents), anal gland smearing, vaginal rubbing, and cloaca rubbing are common (Gleason and Reynierse 1969). In arboreal marsupials (e.g., koalas and phalangers) and New World primates (e.g., marmosets and tree shrews) that contact tree branches while climbing, brachial and sternal scent glands are often evident. Similarly, the passive marking of an environment during the course of other activities is associated with appropriately located scent glands. The interdigital glands of the mule deer (*Odocoileus hemionus*), for example, deposit a secretion on trails whenever the animal moves. This type of passive marking also occurs in species possessing pedal glands (e.g., Cervidae, Suidae, and possibly man). The male mouse, *Mus musculus*, and possibly other rodent species, deposits urine by dipping its penis to the ground as it moves along (Desjardins, Maruniak, and Bronson 1973). This may not be entirely a passive response, as subordinate or castrate males fail to mark in this way. The behavior seems analogous to urine marking in male dogs (B. L. Hart 1974).

2.

For proficient use, a scent gland must be regulated in such a way that secretion is readily available and can be easily applied. This means the sebum must be stored or rapidly synthesized. Moreover, the external ducts must provide openings to the environment that match the manner in which the substratum is approached. For example, the hairs protruding from the ventral gland of the Mongolian gerbil are grooved and orient toward the rear to allow the best possible deposition of sebum as the animal moves forward over an object.

3.

Sexual dimorphism and seasonal variation should be evident in scent glands related to reproduction, implying that hormonal control is essential. Many of the species described in Table 1 show sexually dimorphic scent glands, with the male possessing the larger gland and usually the more frequent marking behavior. As expected, dimorphism of the glands develops at puberty, as in the European rabbit, sugar glider, Mongolian gerbil, and golden hamster. Gland size is attenuated by castration and is exaggerated by the injection of sex hormones, as in the European rabbit, Columbian ground squirrel, Mongolian gerbil, dromedary, golden hamster, and guinea pig.

Likewise, seasonal variation is evident in the glandular activity of many species. Quay (1953) describes variations in five species of *Dipodomys* and states that seasonal differences are apparent. The European rabbit (Mykytowycz 1966*a*; 1966*b*; 1966*c*) and wood rat (*Neotoma fascipes*) (Linsdale and Levis 1951) both show scent gland activity that quickens during the breeding season. Apparently many scent glands and, presumably, related behaviors are most evident during life stages and seasons when sexual communication is necessary (also see Thiessen and Rice 1976).

4.

To be most effective a scent gland should be stimulated by internal or external events most closely related to significant social behaviors. Evidence here is almost entirely lacking, with the obvious exception that the hormone status is critically important. In the female golden hamster, clitoral gland secretion follows examination of the lateral glands by the male (Lipkow 1954). Nichol (1938) reports that male deer discharge scents when frightened or calling fawns. And, according to Mykytowycz (1970), musk glands discharge in fright situations in a wide variety of species (civets [Viverridae] and skunks [Mustelidae]).

5.

A correspondence should exist within a species between the chemical nature of the scent gland secretion and the development of the receptor system. This criterion is one of the most difficult to establish. At the most basic level, it is expected that only macrosmatic animals would possess scent glands and that the olfactory material would be volatile and easily captured by the olfactory epithelium. We must also consider the possibility of interspecific communication in which one species transmits a signal and another receives the chemical message ("allomone"). In this case only the receiver need be macrosmatic. Unfortunately, too little research has been conducted with mammals to establish these relationships, but present information does not support a significant difference in olfactory sensitivity between macrosmatic and microsmatic species (Laing 1975).

The successful isolation and identification of chemical signals may offer important keys to the understanding of complex neurological processes of olfaction. Once the stimulus is clearly defined, a detailed search can begin for olfactory receptors and electrophysiological changes in the central nervous system. Brownlee, Silverstein, Müller-Schwarze, and Singer (1969) have recently identified the compound which elicits aggressive behavior in black-tailed deer (*Odocoileus hemionus*). Michael, Keverne, and Bonsall (1971) have identified a female sex attractant in monkeys, and Ebling (1963), Mykytowycz (1965), and Thiessen (1968) are attempting to isolate olfactory pheromones in monkeys, rabbits, and gerbils, respectively. As noted earlier the glandular products of the rabbit from different regions of the body differ and prove to be complex mixtures of lipids, carbohydrates, and proteins (Goodrich and Mykytowycz 1972). One major component of the ventral scent gland of the Mongolian gerbil is phenylacetic acid (see later sections).

6.

Scent glands should have exaggerated signaling qualities when other sensory systems are not used. Similarly, chemical communication will be enhanced when it is integrated with other modes of communication. Thus, it is expected that olfactory signaling will predominate in (*a*) nocturnal species that cannot rely on visual displays, (*b*) species that live in thick foliage where visual displays are likely to be obscured, (*c*) terrestrial species whose vision or audition is obscured by terrain features, and (*d*) species with short home ranges where olfactory material can provide reliable signposts. Scent glands, in particular, may be used by desert species that must conserve body water and hence cannot utilize desiccating responses such as urination, defecation, or salivation for object marking. In short, scent glands and marking must show ecological adaptiveness. No doubt some species use similar modes of signaling because of phylogenetic relations. But in other cases convergent evolution has driven diverse species toward a common method of communication. It is in the latter cases that ecological adaptation takes precedence over common heritage.

Cercopithecodia species, such as the African and Asian baboons, macaques, guenons, mangabeys, langurs, colobus, and patas monkeys, which are swift arboreal, semiterrestrial, and terrestrial species with large home ranges, do not have organized scent glands or marking patterns. Other primates (e.g., *Pithecia monocha, Cacajao rubicundus*, and *Ateles*) often assume upright postures but have difficulty balancing. Perhaps, as a consequence, the frequency of scent marking is reduced (Moynihan 1967). Man, of course, as an upright, highly mobile species, relies mainly on vision and audition for distance communication; yet olfactory communication may occur in certain cases (McClintock 1971).

Chemical communication becomes most effective when it is combined with other displays. Certain species of lemurs have evolved elaborate combinations of responses for social communication. According to Jolly, *Lemur catta* is exemplary: "A stink-fight is a long series of palmar-marking, tail-marking, and tail-waving directed by two males toward each other. The animals stand between 3 and 10 m apart. First one marks, then the other, with pauses between. Occasionally both tail-wave simultaneously, the two arched backs and tails reflecting each other like a heraldic design. The more aggressive male gradually moves forward, the other retreats, although they are often not close enough to supplant each other in one leap" (Jolly 1966, p. 103).

Obviously no single criterion or set of criteria can predict the appearance and use of scent glands for communication. Many reasons for the presence of scent glands are still unclear. Both diurnal and nocturnal species have marking glands, as do arboreal, semiarboreal, and terrestrial animals. Glands are neither restricted to desert forms nor absolutely linked to burrowing animals. Finally, phylogenetic status in and of itself is not a criterion for scent gland use. More important, no doubt, are the type of ecology faced by the organism and the efficiency of scent marking as a signaling system relative to other potential systems.

8

7.
Last, and most significant, scent marking should be more evident in gregarious species than in others and have an obvious relation to sexual fitness and gene transmission. Pfeiffer (1962), for instance, finds that the most extensive use of pheromones occurs in fish with complex social behaviors. Young (1950) points out a similar distinction between the social frogs, *Bufo*, and the semisocial groups, *Rana*. Likewise, the swamp rabbit (*Sylvilagus aquaticus*) marks by chinning much more frequently than does the cottontail rabbit (*Sylvilagus floridanus*) (Marsden and Holler 1964). The former species is highly territorial and possesses a large chin gland, while the latter species is never observed to defend a territory and has a small chin gland. According to Mykytowycz (1974), similar differences exist in the lagomorphs (*Oryctolagus cuniculus* and *Lepus europaeus*). *Oryctolagus cuniculus* is highly territorial, has large anal marking glands, and marks frequently. *Lepus europaeus*, on the other hand, is a solitary-living hare with a wide home range. Its anal gland, as expected, is exceedingly small. However, there are apparent exceptions. The presence of a marking gland is not always a good indication that a species is sociable; and conversely, the sociality of a species gives no hint that a gland will be found. Eisenberg (1967) points out that solitary, semisocial, and social forms of rodents may or may not show scent marking. Perhaps this latter observation simply reflects our present ignorance about the ecological and social relevance of glandular marking and the use of other sensory modalities.

Ultimately, the importance of a scent gland must be judged by its contribution to the selective fitness of the individual or the population. In one way or another, a functional scent gland has a bearing on individual and species survival. The primary goals of olfactory signaling are set down in Table 3. Of course, not every glandular secretion will have the same function. Species differ in basic needs, and separate glands can serve different purposes. The social and ecological context will specify the function. For instance, phalangers (*Petaurus papuanus*) discriminate individuals and groups on the basis of odor; moreover, they mark territories, themselves, and each other (Schultze-Westrum 1965). At least six forms of scent marking take place, with apparently distinctive secretions from specialized sternal, frontal, and anal glands, and even from saliva. Mykytowycz (1966a) established that European rabbits produce two types of fecal pellets. Some are not marked by anal glands and have no particular territorial significance; others are dampened by anal secretions during excretion and are used to build territorial signposts. Indeed, rats and gerbils can detect differences in odor cues between trails left by animals that have been rewarded or nonrewarded for a response (Davis, Crutchfield, Shaver, and Sullivan 1970; McHose 1967; McHose and Ludvigson 1966; Ludvigson and Sytama, 1964; Topping and Cole 1969). Tavolga (1956) reports that the female gobiid fish (*Bathygobius soporator*) secretes an ovarian material when she is gravid which stimulates courtship in males. Apparently male rhesus monkeys can also distinguish between receptive and nonreceptive females on the basis of hormone-dependent odors (Michael and Zumpe 1970). Obviously, the more we learn about social interactions, the more we will be able to identify scent gland function.

Table 1.
Representative mammalian species with specialized scent glands and behavioral marking

Order	Family	Genus	Species	Common Name	Distribution
Artiodactyla (ungulates)	Bovidae	*Antilope*	*cervicapra*	Black buck	South Asia
	Cervidae	*Capreolus*	*capreolus*	Roe deer	Europe and Asia
	Bovidae	*Cephalophus*	*maxwelli*	Maxwell's duiker	Central West Africa
	Bovidae	*Rupicapra*	*rupicapra*	Chamois	Europe and Asia
	Tayassuidae	*Tayassu*	*tajacu*	Collared peccary	South and Central America
	Cervidae	*Odocoileus*	*hemionus*	Mule deer (black-tailed deer)	Western and North America
	Antilocapridae	*Antilocapra*	*americana*	Pronghorn antelope	North America
Carnivora (carnivores)	Canidae	*Nyctereutes*	*procyonoides*	Racoon dog	East Asia and South America
	Canidae	*Vulpes*	*vulpes*	Red fox	North America, Asia, North Africa, and Europe
	Viverridae	*cryptoprocta*	*ferox*	Fossa	Madagascar
	Viverridae	*Arctictis*	*binturong*	Binturong	South Asia, Java, Sumatra, Philippines, etc.
	Viverridae	*Helogala*	*undulata*	Dwarf mongoose	East Africa
	Viverridae	*Herpestes*	*edwardsi*	Indian grey mongoose	South Asia
Edentata (edentates)	Bradypodidae	*Bradypus*	*tridactylus*	Three-toed sloth	South America
Lagomorpha (rabbits and hares)	Leporidae	*Oryctolagus*	*cuniculus*	European rabbit	Europe and North America
	Leporidae	*Sylvilagus*	*aquaticus*	Swamp rabbit	North America
	Leporidae	*Sylvilagus*	*floridanus*	Eastern cottontail rabbit	North America
	Ochotonidae	*Ochotona*	*princeps*	American pika	North America

Scent Marking, Chemocommunication, Gland Characteristics

Gland Characteristics	Behavioral Characteristics and Social Implications	References
Preorbital gland	Objects marked with preorbital secretion.	Hediger (1949); Tembrock (1968)
Forehead gland	Marks branches at territorial boundaries.	Prior (1968)
Preorbital gland	Objects and conspecifics marked, especially by dominant male. Individuals press glands together. Males mark more frequently, and both sexes mark more frequently in presence of same sex.	Ralls (1971); Pocock (1910)
Occipital gland	Social interaction stimulates marking. Dominant animals mark more frequently than subordinate animals.	Kramer (1970)
Lumbar glandular area	Bends hind legs and rubs gland on grass, tree stumps, and other objects, depositing milky secretion.	Fradrich (1967)
Preorbital gland; tarsal gland; metatarsal gland; interdigital glands	Preorbital gland used to mark objects. Tarsal gland used in aggressive displays by dominant males, and in males and females for individual and sex recognition.	Brownlee et al. (1969); Müller-Schwarze (1967; 1969 a, b; 1971); Müller-Schwarze and Müller-Schwarze (1969)
Subauricular gland	Gland used to mark bushes and grasses, especially around borders of territories. Functions are androgen dependent.	Moy (1970); Müller-Schwarze (1971); Gilbert (1974)
Glands at dorsal root of tail	Glands rubbed on roof of lair.	Ewer (1968)
Glands at dorsal root of tail	Glands rubbed on entrance of refuge.	Ewer (1968)
Chest gland	Gland larger in males, although both males and females mark substratum. Glands maximally active during breeding season.	Ewer (1968); Vosseler (1929)
Perineal gland	Substrate marking, with males marking more frequently.	Kleiman (1974)
Facial glands and chest glands	Marks substratum with chest glands.	Ewer (1968); Tembrock (1968); Zannier (1965)
Facial glands	Marks objects with cheek glands.	Dücker (1965); Ewer (1968)
Mid-dorsal skin gland	Back rubbed against substrate.	Tembrock (1968)
Apocrine chin gland and anal gland	Marking generally done by dominant male. Gland and marking more prominent in male and are androgen dependent.	Myers and Poole (1961); Mykytowycz (1965; 1968); Wales and Ebling (1971)
Chin gland	Marking generally done by dominant male and in the presence of a female, another male, or in isolation. Underside of chin rubbed on objects.	Marsden and Holler (1964)
Chin gland	Similar to swamp rabbit, but marking less frequent.	Marsden and Holler (1964)
Chin gland	Gland rubbed on twigs and other objects.	Harvey and Rosenberg (1960); Barash (1973); Sharp (1973)

11

Order	Family	Genus	Species	Common Name	Distribution
Marsupialia (marsupials)	Dasyuridae	*Antechinus*	*flavipes*	Yellow-footed marsupial mouse	Australia
	Phalangeridae	*Petaurus*	*breviceps*	Sugar glider (short-headed flying pha-langer)	Australia and New Guinea
	Phalangeridae	*Phascolarctus*	*cinereus*	Koala	Australia
	Phalangeridae	*Trichosurus*	*vulpecula*	Brush-tailed phalanger	Australia
Primates (monkeys, apes, and prosimians)	Cebidae	*Ateles*	*geoffroyi*	Central American spider monkey	Central and South America
	Cebidae	*Callicebus*	*moloch*	Orabussu titi	South America
	Cebidae	*Lagothrix*	*cana*	Smoky woolly monkey	South America
	Cebidae	*Lagothrix*	*logothricha*	Humboldt's woolly monkey	South America
	Cebidae	*Saimiri*	*sciureus*	Common squirrel mon-key	South America
	Cebidae	*Callimico*	*goeldii*	Goeldi's mon-key	South America
	Callithricidae	*Callithrix*	*jacchus*	Common mar-moset	South America
	Callithricidae	*Callithrix*	*leucophala*	White-footed marmoset	South America
	Callithricidae	*Leontideus*	*rosalia*	Golden lion marmoset	South America
	Callithricidae	*Callithrix*	*argentatus*	Silvery mar-moset	South America
	Indriidae	*Propithecus*	*verreauxi*	Verreaux's sifaka	Madagascar
	Lemuridae	*Hapalemur*	*griseus*	Grey gentle lemur	Madagascar
	Lemuridae	*Lemur*	*catta*	Ring-tailed lemur	Madagascar
	Lemuridae	*Lemur*	*macaco*	Black lemur	Madagascar
	Lorisidae	*Loris*	*tardigradus*	Slender loris	South Asia
	Tupaiidae	*Tupaia*	*glis*	Common tree shrew	South Asia, Sumatra, Java, Borneo
Proboscidea (elephants)	Elephantidae	*Loxodonta*	*africana*	African ele-phant	Africa

Scent Marking, Chemocommunication, Gland Characteristics

Gland Characteristics	Behavioral Characteristics and Social Implications	References
Sternal glands	Chest rubbing on branches. Glands and behavior especially evident in males.	Ewer (1968)
Frontal glands and sternal glands	Marking of self, mate, and territory, especially by dominant male. Odors of individuals and groups can be distinguished.	Ewer (1968); Schaffer (1940); Schultze-Westrum (1965; 1969); Tembrock (1968)
Sternal gland	Marking more often in male than in female.	Ewer (1968)
Sternal glands	Chest marking of branches most common.	Ewer (1968)
Apocrine glands on sternal, gular, and epigastric areas	Chest marks objects.	Eisenberg and Kuehn (1966); Epple and Lorenz (1967)
Apocrine glands on sternal, gular, and epigastric areas	Chest marks branches and hands, especially in intergroup encounters.	Mason (1966); Moynihan (1966)
Apocrine glands on sternal, gular, and epigastric areas	Chest marks objects.	Epple and Lorenz (1967)
Apocrine glands on sternal, gular, and epigastric areas	Regular marking areas. Chest marks objects.	Epple and Lorenz (1967)
Apocrine glands on sternal, gular, and epigastric areas	Regular marking areas.	Epple and Lorenz (1967)
Sebaceous glands on sternal and gular areas	Regular marking areas. Marking glands mature at puberty; marking is done mostly by dominant male and is directed toward objects or conspecifics.	Epple and Lorenz (1967)
Sebaceous glands on sternal and gular areas.	Marking in dominant male increases after fighting encounters but not in female.	Epple and Lorenz (1967); Epple (1974)
Sebaceous glands on sternal and gular areas	See comments for *Callimico goeldii*.	Epple (1967)
Sebaceous glands on sternal and gular areas	Regular marking areas. (See comments for *Callimico goeldii*.)	Epple and Lorenz (1967)
Sebaceous glands on sternal and gular areas	See comments for *Callimico goeldii*.	Epple (1967)
Long scent gland on ventral surface of throat	Throat marking most frequent where females urinate and during intertroop encounter.	Jolly (1966); Petter (1962a, b)
Brachial gland	Marking has been reported.	Andrew (1964)
Sebaceous brachial gland on upper chest near armpit; apocrine and eccrine glands on forearm	Complex fighting display in which glands are used to anoint the body, especially the tail, which is waved at other animals.	Jolly (1966); Petter (1965); Petter-Rousseaux (1964); Evans and Goy (1968)
Sebaceous glands in anogenital region	Males mark females and objects. Dominant males will mark subordinate males.	Jolly (1966); Petter (1962a, b; 1965)
Apocrine brachial gland	Marking known but not described in detail.	Hill (1956a, b)
Apocrine glands from chin onto sternum; larger in male	Regular marking areas. Marking and gland more pronounced in dominant male, especially during aggressive encounters.	Andrew (1964); Kaufmann (1965); Sprankel (1961)
Apocrine temporal gland	Scent delivered to conspecific by trunk. Gland most active during rut.	Kuhme (1961); Tembrock (1968)

Order	Family	Genus	Species	Common Name	Distribution
Rodentia (rodents)	Cricetidae	*Arvicola*	*terrestris*	Water vole	Europe and Asia
	Cricetidae	*Meriones*	*hurrianae*	Gerbil	North Africa and Asia
	Cricetidae	*Meriones*	*unguiculatus*	Mongolian gerbil	Northeast Asia
	Cricetidae	*Meriones*	*tristrami*	Gerbil	North Africa
	Cricetidae	*Mesocricetus*	*auratus*	Golden hamster	East Europe and West Asia
	Cricetidae	*Neotoma*	*fascipes*	Wood rat	North and Central America
	Heteromyidae	*Dipodomys*	*nitratoides*	Kangaroo rat	North America
	Heteromyidae	*Heteromys*	*anomalas*	Spiny pocket mouse	Central and South America
	Heteromyidae	*Heteromys*	*iepturus*	Spiny pocket mouse	Central and South America
	Heteromyidae	*Liomys*	*pictus*	Spiny pocket mouse	Central America
	Muridae	*Rattus*	*rattus*	Black rat	Europe, Asia, Australia, and Africa
	Muridae	*Cricetomys*	*gambianus*	African giant rat	Africa
	Sciuridae	*Citellus*	*beecheyi*	California ground squirrel	East Europe, Asia, and North America
	Sciuridae	*Spermophilus*	*columbianus*	Columbian ground squirrel	North America

14

Scent Marking, Chemocommunication, Gland Characteristics

Gland Characteristics	Behavioral Characteristics and Social Implications	References
Flank glands	Marking accomplished by rapid wiping of flanks by hind feet, followed by foot-stomping.	L. E. Brown (1966); Frank (1956)
Ventral scent gland	Ventrum rubbed on surfaces. Response often associated with "sandbathing" movements.	Eisenberg (1967)
Sebaceous ventral gland and gular sebaceous gland	Marking generally by ventral gland, especially in dominant males. Gland and marking more prominent in male and are androgen dependent in male. Defeated males avoid smell of dominant animals.	Eisenberg (1967); Nyby, Thiessen, and Wallace (1970); Thiessen (1968)
Sebaceous ventral gland	Marking by ventral gland especially in males. Gland and marking are androgen dependent.	Thiessen, Wallace, and Yahr (1973)
Sebaceous gland on flanks	Regular marking areas. Marking prominent around diestrous females or other hamster smells. Male will mark defeated rival, who in turn will subsequently avoid smell of dominant animal. Gland and marking appear at puberty and are androgen dependent. Marking more frequent in male, but both sexes mark more frequently in presence of same sex.	Dieterlen (1959); Eibl-Eibesfeldt (1953 a, b, c, d); Ewer (1968); Johnston (1969; 1970); Murphy (1970 a, b, 1971); Johnston (1975 a, b, c); Drickamer and Vandenbergh (1973); Drickamer, Vandenbergh, and Colby (1973)
Ventral scent gland	Ventral marking of rocks, logs, etc., presumably for the purpose of defining territories.	Bailey (1936); Howell (1926); Linsdale and Levis (1951); Werner, Dalquest, and Roberts (1952)
Sebaceous dorsal gland	Regular marking area. Marking increases during exploration of novel area. Gland is apparently androgen specific. Ventral rub integrated with side-rubbing movements.	Eisenberg (1963; 1967); Quay (1953)
Ventral scent gland	Ventrum rubbed on surfaces, usually in preferred areas. Not integrated with side-rubbing movements.	Eisenberg (1963; 1967)
Ventral scent gland	Ventrum rubbed on surfaces, usually in preferred areas. Not integrated with side-rubbing movements.	Eisenberg (1963; 1967)
Ventral scent gland	Ventrum rubbed on surfaces, usually in preferred areas. Not integrated with side-rubbing movements.	Eisenberg (1963; 1967)
Sebaceous ventral gland	Regular marking area. Marking intense following exploration of novel area. Ventral rub integrated with side-rubbing movements.	Eisenberg (1963; 1967)
Cheek gland	Males, especially, mark substrate with cheeks.	Ewer (1967)
No apparent specialized gland area	Males and females mark objects by applying the entire ventral surface. Marking occurs especially during disturbances and entry of strange animals.	Ewer (1961); Linsdale (1946)
Dorsal skin gland and mouth gland	Animals mark by turning over to press gland against object or by rubbing gland against overhang. Marking and glandular integrity are androgen dependent.	Kivett (1975); Steiner (1973)

Order	Family	Genus	Species	Common Name	Distribution
	Sciuridae	*Spermophilus*	*undulatus*	Arctic ground squirrel	North America
	Sciuridae	*Marmota*	*olympus*	Olympic marmot	Europe, Asia, and North America
	Caviidae	*Geocapromys*	*ingrahami*	Bahaman hutia	West Indies

Scent Marking, Chemocommunication, Gland Characteristics

Gland Characteristics	Behavioral Characteristics and Social Implications	References
Dorsal skin gland and mouth gland	Animals exchange scent from mouth gland with a greeting kiss. They also mark the substrate with cheeks, throat, and belly. Males mark more often than females.	Steiner (1973)
Dorsal skin gland and mouth gland	Animals exchange scents from mouth gland with a greeting kiss. They also mark the substrate with cheeks, throat and belly. Males mark more often than females.	Barash (1973)
Perineal gland	Substrate marking.	Howe (1974)

Table 2.
Marking characteristics of various species

Order	Family	Genus	Species

Anal Gland Marking

Order	Family	Genus	Species
Carnivora	Mustelidae	*Eira*	*barbara*
	Mustelidae	*Martes*	*martes*
	Mustelidae	*Meles*	*meles*
	Mustelidae	*Mellivora*	*capensis*
	Mustelidae	*Mephitis*	*mephitis*
	Mustelidae	*Mustela*	*putorius*
	Viverridae	*Atilax*	*pauludinosus*
	Viverridae	*Crossarchus*	*obscurus*
	Viverridae	*Helogala*	*undulata*
	Viverridae	*Herpestes*	*edwardsi*
	Viverridae	*Mungos*	*mungo*
	Viverridae	*Suricata*	*suricata*
Insectivora	Solenodontidae	*Solenodon*	*paradoxus*
Lagomorpha	Leporidae	*Oryctolagus*	*cuniculus*
Marsupialia	Phalangeridae	*Trichosurus*	*vulpecula*
Primates	Daubentoniidae	*Daubentonia*	*madagascariensis*
	Lemuridae	*Lemur*	*catta*
	Lemuridae	*Lemur*	*fulvus*
	Lemuridae	*Lemur*	*macaco*
	Lemuridae	*Microcebus*	*murinus*
	Lemuridae	*Microcebus*	*coquereli*
Rodentia	Caviidae	*Cavia*	*porcellus*
	Cricetidae	*Ondatra*	*zibetica*
	Dasyproctidae	*Dasyprocta*	*aguti*
	Dasyproctidae	*Myoprocta*	(two species)
	Gliridae	*Glis*	*glis*
	Sciuridae	*Marmota*	*marmota*

18

Scent Marking, Chemocommunication, Gland Characteristics

Common Name	Distribution	Reference
Tayra	Central and South America	Kaufmann and Kaufmann (1965)
Pine marten	Europe and Asia	Goethe (1938; 1940)
Eurasian badger	Europe and Asia	Eibl-Eibesfeldt (1950)
Ratel	Africa and South America	Sikes (1964)
Striped skunk	North America	Bourlière (1970)
European polecat	Europe, Asia, and North America	Goethe (1938; 1940)
Marsh mongoose	Africa	Dücker (1965)
Kusimanse	Central West Africa	Dücker (1965)
Dwarf mongoose	Africa	Zannier (1965)
Indian grey mongoose	South Asia	Hediger (1955)
Banded mongoose	Africa	Dücker (1965)
Slender-tailed meerkat	South Africa	Ewer (1968)
Haitian solenodon	Haiti	Andrew (1964)
European rabbit	Europe and North America	Mykytowycz (1966 a, b, c; 1970)
Brush-tailed phalanger	Australia	J. A. Thompson and Pears (1962)
Aye-aye	Madagascar	Petter (1965)
Ring-tailed lemur	Madagascar	Andrew (1964); Evans and Goy (1968)
Brown lemur	Madagascar	Petter-Rousseaux (1964)
Black lemur	Madagascar	Petter (1965)
Lesser mouse lemur	Madagascar	Petter (1965)
Coquerel's mouse lemur	Madagascar	Petter (1965)
Guinea pig	South America	Kunkel and Kunkel (1964)
Muskrat	North America	Darchen (1964)
Agouti	Central and South America	Roth-Kolar (1957)
Acouchi	South America	Morris (1962)
Fat dormouse	Europe and Asia	Koenig (1960)
Marmot	Europe, Asia, and North America	Müller-Using (1956)

Order	Family	Genus	Species
	Sciuridae	*Citellus*	*richardsonii*

Urination (Including Urine Washing)

Order	Family	Genus	Species
Artiodactyla	Cervidae	*Cervus*	*canadensis*
	Cervidae	*Odocoileus*	*hemionus*
	Cervidae	*Rangifer*	*tarandus*
	Hippopotamidae	*Choeropsis*	*liberiensis*
Carnivora	Canidae	*Canis*	*familiaris*
	Canidae	*Canis*	*lupus*
	Felidae	*Felis*	*catus*
	Procyonidae	*Nasua*	*nasua*
Primates	Cebidae	*Cebus*	*apella*
	Cebidae	*Saimiri*	*sciureus*
	Indriidae	*Propithecus*	*verreauxi*
	Lemuridae	*Lemur*	*macaco*
	Lemuridae	*Microcebus*	*murinus*
	Lorisidae	*Galago*	*crassicaudatus*
	Lorisidae	*Galago*	*senegalensis*
	Lorisidae	*Loris*	*tardigradus*
	Pongidae	*Pongo*	*pygmaeus*
	Callithricidae	*Callithrix*	*jacchus*
	Tupaiidae	*Tupaia*	(several species)
Rodentia	Muridae	*Mus*	(several species)
	Muridae	*Rattus*	(several species)
	Muridae	*Cricetomys*	*gambianus*

Scent Marking, Chemocommunication, Gland Characteristics

Common Name	Distribution	Reference
Ground squirrel	East Europe, Asia, and North America	Sleggs (1926)
Roosevelt elk	North America	Graf (1956)
Mule deer (black-tailed deer)	Western North America	Graf (1956)
Reindeer	North Europe, North Asia, Northern North America, and Greenland	Epsmark (1964a, b)
Pygmy hippo-potamus	Central West Africa	Hediger (1950)
Dog (domestic)	Varied	B. L. Hart and Haugen (1971); J. P. Scott and Fuller (1965)
Wolf	Europe, Asia, North America, and Arctic	J. P. Scott and Fuller (1965)
Cat (domestic)	Varied	B. L. Hart (1973); Leyhausen and Wolff (1959)
Coati	Central and South America	Balph and Stokes (1963)
Brown capuchin	South America	Hill (1938; 1960); Nolte (1958)
Common squirrel monkey	South America	Castell and Maurus (1967); Hill (1938; 1960)
Verreaux's sifaka	Madagascar	Jolly (1966)
Black lemur	Madagascar	Petter (1962a, b); Jolly (1966)
Lesser mouse lemur	Madagascar	Petter (1965)
Thick-tailed bushbaby	Africa	Andrew (1964)
Senegal	Africa	Doyle, Pelletier, and Bekker (1967)
Slender loris	South Asia	Hill (1938); Ilse (1955)
Orangutan	Sumatra and Borneo	Davenport (1967)
Common marmoset	South America	Epple (1970)
Tree shrews	South Asia, Sumatra, Java, and Borneo	Sprankel (1961); Kaufmann (1965)
House mouse	Varied	Tembrock (1968); Jones and Nowell (1973); Desjardins, Maruniak, and Bronson (1973)
Rat	Varied	Tembrock (1968)
African giant rat	Africa	Ewer (1967)

Order	Family	Genus	Species
	Sciuridae	*Sciurus*	(several species)
	Cricetidae	*Peromyscus*	*maniculatus*
	Caviidae	*Geocapromys*	*ingrahami*

Defecation

Order	Family	Genus	Species
Artiodactyla	Hippopotamidae	*Choeropsis*	*liberiensis*
Lagomorpha	Leporidae	*Oryctolagus*	*cuniculus*
Perissodactyla	Rhinocerotidae	*Rhinoceros*	*unicornis*
Primates	Lemuridae	*Cheirogaleus*	(several species)
	Pongidae	*Pongo*	*pygmaeus*
Rodentia	Muridae	*Mus*	(several species)
	Muridae	*Rattus*	(several species)

Cloaca Rubbing

Order	Family	Genus	Species
Marsupialia	Dasyuridae	*Antechinus*	*flavipes*
	Dasyuridae	*Dasycercus*	*cristicauda*
	Dasyuridae	*Sarcophilus*	*harrisi*
	Dasyuridae	*Sminthopsis*	*crassicaudata*
	Phalangeridae	*Petaurus*	*breviceps*
Monotremata	Tachyglossidae	*Tachglossus*	*aculeatus*

Miscellaneous Marking

Order	Family	Genus	Species
Artiodactyla	Bovidae	*Tragelaphus*	(several species)[a]
	Camelidae	*Camelus*	*dromedarius*[b]

Scent Marking, Chemocommunication, Gland Characteristics

Common Name	Distribution	Reference
Tree squirrels	Europe, Asia, North and South America	Tembrock (1968)
Deer mouse	North and Central America	Eisenberg (1963)
Bahaman hutia	West Indies	Howe (1974)
Pygmy hippo-potamus	Central West Africa	Hediger (1950)
European rabbit	Europe and North America	Mykytowycz (1966 *a*, *b*, *c*; 1970)
Great Indian rhinoceros	South Asia	Gee (1953)
Dwarf lemur	Madagascar	Petter (1965); Petter and Petter-Rousseaux (1956)
Orangutan	Sumatra and Borneo	Davenport (1967)
House mouse	Varied	Tembrock (1968)
Rat	Varied	Tembrock (1968)
Yellow-footed marsupial mouse	Australia	Ewer (1968)
Crest-tailed marsupial mouse	Australia	Ewer (1968)
Tasmanian devil	Tasmania	Ewer (1968)
Fat-tailed sminthopsis	Australia	Ewer (1968)
Sugar glider (short-headed flying phalanger)	Australia and New Guinea	Schultze-Westrum (1965)
Australian echidna	Australia	Dobroruka (1960); Hediger and Kummer (1956)
Spiral-horned antelope	Africa	Walther (1964)
Arabian camel	North Africa and Southwest Asia	Pilters (1956)

Order	Family	Genus	Species
	Cervidae	*Muntiacus*	*muntjak* [c]
Carnivora	Felidae	*Felis*	*catus* [a]
	Viverridae	*Helogala*	*undulata* [a]
	Viverridae	*Herpestes*	*edwardsi* [a]
Insectivora	Tenrecidae	*Echinops*	*telfairi* [d]
Marsupialia	Dasyuridae	*Antechinus*	*flavipes* [d]
	Dasyuridae	*Dasycercus*	*cristicauda* [d]
	Dasyuridae	*Sminthopsis*	*crassicaudata* [d]
	Phalangeridae	*Petaurus*	*breviceps* [d]
Primates	Callithricidae	*Callithrix*	*jacchus* [e]
	Cebidae	*Callicebus*	*moloch* [a]
	Indriidae	*Propithecus*	*verreauxi* [e]
	Lemuridae	*Lemur*	*catta* [e]
	Lemuridae	*Lemur*	*macaco* [e]
Rodentia	Muridae	*Mus*	(several species) [c]

[a] *Cheek marking.*
[b] *Neck marking.*
[c] *Foot marking.*
[d] *Saliva marking.*
[e] *Vaginal marking.*

Scent Marking, Chemocommunication, Gland Characteristics

Common Name	Distribution	Reference
Indian muntjak	South Asia, Sumatra, Java, and Borneo	Tembrock (1968)
Cat (domestic)	Varied	Ewer (1968)
Dwarf mongoose	Africa	Zannier (1965)
Indian grey mongoose	South Asia	Dücker (1965)
Tenrec	Madagascar	Eibl-Eibesfeldt (1965)
Yellow-footed marsupial mouse	Australia	Ewer (1968)
Crest-tailed marsupial mouse	Australia	Ewer (1968)
Fat-tailed sminthopsis	Australia	Ewer (1968)
Sugar glider (short-headed flying phalanger)	Australia and New Guinea	Schultze-Westrum (1965)
Common marmoset	South America	Epple (1970)
Orabussu titi	South America	Moynihan (1966)
Verreaux's sifaka	Madagascar	Jolly (1966)
Ring-tailed lemur	Madagascar	Petter (1965)
Black lemur	Madagascar	Petter (1965)
House mouse	Varied	Tembrock (1968)

Table 3.
Integrative functions of scent gland and olfactory communication

Signal Function	Significance for the Gene Pool
Aggregation	Signals may have directional qualities, bringing conspecifics together so that territories or dominance orders can be established and differential mating occur. Aggregation is primary to any other function and may be fundamentally important for individual and group survival.
Recognition	In restricted gene pools it is essential that social and physiological distinctions can be made with ease. Animals must be able to discriminate individuals, group and species memberships, age, sex, social status among peers, and reproductive and aggressive status of other males and females. Territories, home ranges, nest areas, and other locations must be recognized with ease.
Territorial personalization, maintaining social status, and causing dispersal	Territories or social status are assumed and defended for multiple reasons. These behaviors personalize situations, provide food, shelter, and mates, disperse population members, encourage migration, reduce overt aggression and the spread of disease and parasites, decrease predation, increase inbreeding, and maintain reproductive potential among nonbreeding subordinates. In short, personalization with scent signals increases security.

General Features of Scent Glands and Sebaceous Tissue

The following discussion of the nature of scent glands is based on reports by Doty and Kart (1972), Ebling (1963), Ebling, Elling, Skinner, and White (1970), Glenn and Gray (1964), Mitchell (1965), Montagna (1962), Nikkari and Valavaara (1969), Strauss and Ebling (1970), Strauss and Pochi (1963), Thoday and Shuster (1970), and research from our laboratory.

Table 1, which lists glandular characteristics associated with scent marking, presents few distinctions between holocrine and apocrine glands. Often the discrimination is not recognized by investigators or confusion exists over terminology. For instance, it is common to encounter a division between "sebaceous" and "apocrine" glands; however, we believe it is more correct to specify that the term *sebaceous* refers to a particular type of cell exuding fatty materials (i.e., sebum), while *holocrine* and *apocrine* refer to ways in which sebum is formed. At the moment, there is no reason to believe that scent marking is differentially related to holocrine and apocrine glands, although this possibility still exists.

The development of sebaceous glands and their bodily distribution are intimately associated with hair follicles in most mammals. In man and other primates, sebaceous tissue is widespread, appearing most commonly in the scalp, forehead, face, chin, and genital regions. In man, somewhere between four-hundred and nine-hundred glandular cells may be counted in each square centimeter of skin surface.

To the extent that studies permit generalizations, it appears that many species which have evolved large aggregates of secretory cells (scent glands) have also evolved complex pheromone systems for communication. These include the species listed in Table 1, as well as mice (*Mus musculus*) and rats (*Rattus norvegicus*), in which the major sebaceous tissue has developed into an internal preputial gland with connections serving the urinary tract.

Sebum generally contains histologically demonstrable cholesterol esters, some phospholipids, and possibly triglycerides. The exact constitution of lipids and associated enzymes is largely unknown but will probably vary widely with species. Civet from the civet cat (*Viverra civetta*) and muskone from the musk deer (*Moschus moschiferus*) are ketones. Prior to 1969, no one had fractionated a complex proteolipid into pheromone-active and pheromone-inactive materials. In that year, however, Brownlee, Silverstein, Müller-Schwarze, and Singer isolated and identified the compound *cis*-4-hydroxydodec-6-enoic acid lactone as the chief pheromone of the tarsal scent gland in the black-tailed deer (*Odocoileus hemionus*). This substance elicits a complex behavioral response related to individual recognition, aggregation, and aggression (see also Müller-Schwarze 1969*a*; 1971). The active principle in boar urine which stimulates the adoption of the mating stance in females is thought to be the steroid 5 α-androst-16-ene-3-one, the source of boar taint in pork (see Davidson and Levine 1972). Recently Michael, Keverne, and Bonsall (1971) have identified vaginal secretions (copulins) in the female rhesus monkey which are estrogen dependent and stimulate copulation by the male. The stimulating substances have been identified by gas chromatography as acetic, propionic, isobutyric, butyric, isovaleric, and isocaproic acids. These organic acids, when artificially combined in the proportions found in the vaginal smears, elicit the male sexual response. It is clear that much more effort is required in the analysis of mammalian pheromones before a level of sophistication can be reached equaling that for insect investigations (Wilson 1966; Butler 1966).

It is generally agreed that sebaceous glands are not innervated but are, rather, hormonally regulated. Although tropic hormone control varies according to species, age, sex, nutrition, social status, ecological features, and other conditions, it is interesting that the typical hormone regulators are the same ones that are responsible for territorial marking, aggression, reproduction, and maternal care. It is as if evolution has stipulated that the entire collage of related functions be controlled by a similar hormone balance—certainly a parsimonious arrangement to insure gene transmission.

In mice and rats, testosterone promotes proliferation, enlargement, and turnover of sebaceous cells. Removal of the pituitary, however, attenuates the androgen effect, suggesting a synergism between pituitary products and gonadal androgens. Testosterone apparently acts directly on the sebaceous tissue, since topical application produces gland enlargement.

Estrogen, in contrast to testosterone, decreases sebaceous activity in some species (e.g., man, rabbits, and deer mice, but not the Mongolian gerbil), while at the same time increasing mitotic activity. In large amounts, estrogen will antagonize testosterone when both are administered systematically. No antagonism is evident with topical application of these hormones, suggesting that the opposition is indirect, perhaps involving inhibition of gonadotropic hormones. In mice and rats, glandular activity and cell turnover are tied to the ovarian estrous cycle, showing maxima at early proestrus and minima at estrus. Parenthetically, there is sometimes a negative correlation between mitotic incidence and gland size.

Progesterone is an ineffective stimulus for gland growth in rats and humans in small doses (0.5 mg/day) but will lead to a pronounced enlargement of sebaceous glands in rats receiving larger amounts (10 mg/day). The same is true for the Mongolian gerbil. Progesterone, it must be remembered, is slightly androgenic; moreover, it is easily converted to androgen. Nevertheless, a synthetic progestogen, 17-d-ethynyl-10-nortestosterone (Norlutin), with only three percent androgenic activity, when assayed on seminal vesicles and ventral prostate, enlarges sebaceous glands in prepubertal male rats and causes pseudohermaphroditism in human female infants when given to mothers during pregnancy. Physiological amounts of progesterone given prepubertally to either sex are ineffective, however, and the production of sebum in women shows no increase with the luteal stage of the ovarian cycle when progesterone titers are high.

A critical observation is that in almost all species investigated, including several of those listed in Table 1 and humans, the male has the larger and more active sebaceous glands. The onset of activity is correlated with the onset of puberty or seasonal breeding and can be depressed by castration and elevated with supplemental injections of testosterone. These data and those mentioned earlier are the best evidence to date that sebaceous tissue is controlled primarily by circulating androgens, with supporting influences from the pituitary and modifying effects from estrogens and progestogens.

The Scent Gland of the Mongolian Gerbil

The midabdominal sebaceous gland pad of the Mongolian gerbil, first described by Webster (1963) and Dambach (1964), is holocrine in nature. Figure 1 shows its fusiform taper and the size characteristic of adult males and females. A similar dimorphism in ventral gland size has been recorded for *M. tristrami, M. tamariscinus, M. erythrourus, M. persicus*, and *M. meridianus* (Sokolov and Skurat 1966), and is evident in *M. libycus* and *M. shawi* in our laboratory. No systematic observations of territorial marking have been made on the majority of these species. *M. libycus, M. tristrami*, and *M. shawi* have been observed to scent mark in our laboratory.

The midventral sebaceous pad is the major site, but not the only site, for organized sebaceous tissue in the Mongolian gerbil. The chin and neck areas are also speckled with glandular tissue, though the tissue is not resolved into a gland pad. Importantly, the adult gerbil will occasionally chin objects which are too high to rub with the ventral gland (Thiessen, Yahr, and Lindzey 1971). Perhaps a chin scent is deposited by this behavior. Beyond these observations, little is known about this chin gland complex, except that it seems unresponsive to supplemental injections of testosterone. *M. libycus* and *M. shawi* have elaborated chin glands, but scent marking with these glands has not been observed.

According to Glenn and Gray (1964), the ventral scent gland first appears during development at approximately four weeks of age in the male and sixteen weeks in the female. We have observed specks of sebaceous tissue on the abdominal area within the first few days postpartum. In any case, sexual dimorphism is evident at a very early age and is maintained throughout adulthood (see also Mitchell 1965).

The gland pad is easily exposed by clipping the surrounding hairs, and sebum is then readily visible as an orangish waxy substance. The sebum feels oily and smells musky to humans. As much as six milligrams of sebum can be removed from a male's gland by rubbing it several times with a filter paper strip dampened with 70 percent alcohol. The gland can be effectively cleaned in this manner and the momentary secretory output assessed by weighing the amount of sebum obtained. A correlated measure of sebaceous gland activity can be obtained by removing and weighing the gland pad.

The cleaned gland has a multipore appearance. Glenn and Gray (1964) indicate that several hairs arise from each pore. The hairs terminate subcutaneously, adjacent to secretory units, and resemble miniature troughs. The histological characteristics of the sebaceous gland are beautifully shown in reports by Glenn and Gray (1964) and Mitchell (1965). Little is known about the chemical constitution of the gland, although Glenn and Gray give evidence that free fatty acids and cholesterol are common constituents. Arluk (1968) analyzed the lipid classes of the sebum, using thin-layer and gas chromatography, and concluded that the sebum contains 79.3 percent neutral fats and 10.5 percent free fatty acids. A large proportion of the free fatty acid was tentatively identified as a highly branched, monounsaturated fatty acid of 22 carbon number. A high proportion of the neutral fats (60 percent) were free (unesterified) sterols. Injecting ^{3}H-testosterone twenty-four hours before analysis showed that radioactivity was localized within the upper regions of the alveoli and in the sebum itself. This is evidence that the principal site of testosterone action in the ventral gland is at the cellular level. Our efforts to isolate a pheromone and characterize the sebum will be described in later chapters.

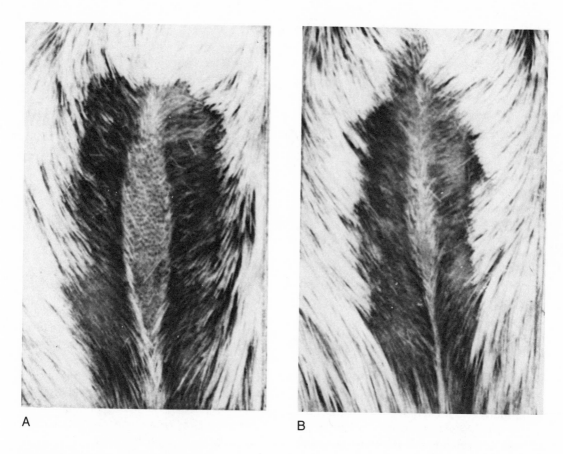

A   B

Figure 1.
Gland characteristics of intact male (*A*) and female (*B*) gerbil.

Other evidence tends to confirm that androgens act on sebaceous tissue at the cellular level. Neumann (1971–1972) has demonstrated that the antiandrogen cyproterone acetate (5.0 mg/day), which competes with androgens at the receptor level, substantially reduces the stimulating effects of TP (testosterone propionate) when given to castrates. Cyproterone acetate prevents the full development of glandular tissue.

The hormone control of gerbil sebaceous glands is similar to, but not identical with, that of mice and rats. No innervation of the gland is obvious, as the gland retains its secretory status when transplanted to the dorsal surface of the animal (Mitchell 1965). Testosterone, in the male, and estrogen and progesterone, in the female, are the normal hormones that regulate the gland size and secretory level. However, a variety of steroids reinstate the gland of castrates of one or both sexes. Among these are estradiol-17-β, testosterone, androstenedione, 1 α-methyltestosterone, 6 α-fluorotestosterone, androstenediol, estrone, estriol, and 19-noretestosterone. More will be mentioned about steroids later with regard to scent marking. In males, maturation of the gland can be prevented by castration at thirty days of age and can be facilitated by weekly injections of 640 micrograms of TP. In adult males, a close relationship exists between hormone titer and gland size (Blum and Thiessen 1971).

Interestingly, the developmental work and the work with adults suggest that there are upper limits to gland growth in males which cannot be exceeded no matter how much testosterone is administered. This contrasts with territorial marking, which can be elevated considerably above control levels (see later chapters). Sexual dimorphism of gland size is apparent regardless of the amount of hormone administered as long as both sexes receive the same amount of hormone (Thiessen, Yahr, and Lindzey 1971).

Glenn and Gray (1964), using immature female gerbils as assay animals, show comparable sensitivity of gland weight to daily administration of TP. Between 40 and 640 micrograms a day for fourteen days elicits a dose-dependent response that is greater than that for any other hormone tested. Both free fatty acids and cholesterol parallel increases in gland weight, but neither offers a better assay measure than gland weight. The investigators reason, therefore, that hormone stimulation does not directly result in activation of enzymes related to glandular sebum constituents. Rather, the growth effects may occur at a more basic level of cellular activity, perhaps involving amino acid incorporation into protein, or DNA and RNA synthesis. This seems to be the case with the flank gland of the golden hamster (Giegel, Stolfi, Weinstein, and Frost 1971), in which castration decreases total DNA and RNA and the ratio of RNA to DNA, and replacement therapy with TP elevates thymidine and uridine uptake.

Glenn and Gray (1964) found significant gland weight responses in immature animals to estrogen (estradiol-17β) in daily doses of 0.2 to 6.4 micrograms over fourteen days. Also, a response to progesterone occurred for daily doses ranging from 50 to 3,200 micrograms. Interestingly, the lowest dose of each hormone gave a response; yet higher doses were not necessarily more stimulating. An upper response level is thereby indicated. Pregnant mare serum (PMS) (9.5–4 IU/day: 14 days), which acts as an ovarian follicular stimulant, also augments gland growth. The response in the female parallels uterine growth. For both males and females the gonads are necessary for the PMS effect, suggesting that the gonadal steroids are important mediators of PMS-induced gland growth.

The augmenting effects of estrogen and progesterone reported by Glenn and Gray for immature gerbils and the stimulating effects of estrogen in intact males observed in our laboratory are in contrast to data reported for mice and rats. In these species, progesterone has a moderate effect but only in large doses, and estrogen may actually inhibit glandular activity. Also in opposition to effects reported for rats, Glenn and Gray found no sebaceous gland response to ACTH (8.0 IU/day: 14 days) or hydrocortisone (1.0 mg/day: 14 days).

In summary, it is apparent that the sebaceous scent pad of the Mongolian gerbil depends primarily on androgen secretion from the gonads, although estrogen and progesterone may have inducing effects. Arluk has found similar hormonal effects on gland size (1966). Estrogen and possibly progesterone are especially important in lactating females, as will be pointed out later. Neural input to the gland is unnecessary for its activity. Developmental responses and sexual dimorphism in adults implicate gonadal maturation as the significant functional determinant. Sebaceous gland growth has obvious limitations under hormone stimulation; nevertheless, both sexes of immature animals and even adults are responsive. Castration and TP replacement therapy produce predicted effects on the gland morphology and function in a dose-dependent manner. ACTH and hydrocortisone do not stimulate gland growth. In many respects the sebaceous gland of the Mongolian gerbil has the same hormone control as do sebaceous tissues in rats, mice, humans, and probably several other species. The major difference is that the gerbil has evolved a complicated chemical communication system dependent on an organized gland pad and associated marking behavior.

The Harderian gland, which is proving to be of importance in the regulation of rodent communication, is described in regard to its function in Chapter 6.

2

General Characteristics of the Mongolian Gerbil

Classification and Origin

Gerbils are in the order Rodentia, the suborder Myomorpha, the family Cricetidae, and the subfamily Gerbillinae (Morris 1964). All animals of the subfamily Gerbillinae are called gerbils; however, these are more appropriately classified by genera and species.

According to Schwentker (1963; 1968) gerbils are divided into ten or possibly twelve genera. A somewhat more detailed classification of genera and species by Morris (1964) is given in Table 4. A detailed description of the taxonomy of the genus *Meriones* is given by Chaworth-Musters and Ellerman (1947).

The genus *Meriones* was first described by Illiger in 1811, and the Mongolian gerbil (*Meriones unguiculatus*) was first identified by Milne-Edwards in 1867. *Meriones* comes from a Greek derivation for "warrior," while *unguiculatus* is Latin for "fingernail." The common name, *gerbil*, is of Arabic origin and originally referred to small jumping rodents in Africa and Asia now referred to as jerboa. The Arabic name, *yarbu*, was translated into Latin as *gerbo*, into French as *gerbille*, and finally into English as *gerbil* (Robinson 1969).

The wide representation of the gerbil in several genera and its relationship to other members of the family Cricetidae enhance its importance in comparative psychological research. Cricetidae is the family most represented in laboratory and field studies of behavior: for example, by the white-footed mice (*Peromyscus*; 57 species), cotton rats (*Sigmodon*; 14 species), wood rats (*Neotoma*; 20 species), golden hamsters (*Mesocricetus auratus*), lemmings (*Lemmus*; 4 species) and common field voles (*Microtus*; 42 species). Indeed, behavioral work may clear up taxonomic difficulties arising out of morphological studies of these species (Eisenberg 1967). Unless otherwise specified, the term *gerbil* in this monograph will refer to the Mongolian gerbil (*M. unguiculatus*).

The Mongolian gerbil is a native of eastern Mongolia, northeast China, and western Manchuria. Tanimoto (1943) lists a number of provinces to which the gerbil is indigeneous; however, these geographical areas are difficult to locate on current maps. As specifically as possible, the species is located in an area west of a line drawn between Harbin and Moukden and north of Ch'engte in Jehol province (approximately 200 km north of Peking). The northern and western boundaries of its range are not clearly defined.

Victor Schwentker of the West Foundation introduced the gerbil into the United States in 1954. He obtained eleven pairs from Michie Nomura of the Central Laboratory for Experimental Animals in Japan. Nomura's original population was obtained from twenty caught by C. Kasuga in the basin of the Amur River in eastern Mongolia in 1935 (Rich 1968). Of the eleven pairs reaching the United States, five females

Table 4.
Genera and species of gerbils

Genus	Species	Common Name	Location
Monodia	*mauritaniae*	African gerbil	Africa
Gerbillus	(37 species)	Smaller gerbil	Southwest Asia and Africa
Tatera	(32 species)	Larger gerbil	South Asia and Africa
Taterillus	(16 species)	Naked-soled gerbil	Africa
Desmodillus	*auricularis*	Cape short-tailed gerbil	South Africa
Desmodilliscus	*braueri*	Short-eared rat	Central Africa
Pachyuromys	*duprasi*	Fat-tailed gerbil	North Africa
Ammodillus	*imbellis*	African gerbil	East Africa
Meriones	(13 species)	Jird	North Africa and Central Asia
Brachiones	*przewalskii*	Przewalski's gerbil	Central Asia
Psammomys	*obesus*	Fat sand rat	North Africa and Southwest Asia
Rhombomys	*opimus*	Great gerbil	Asia

and four males were induced to breed. Progeny of these matings served as the historical nucleus for most research colonies now extant in the United States. Schwentker, at Tumblebrook Farms in Brant Lake, New York, was the major supplier of research animals until the stock was moved to West Brookfield, Massachusetts, by D. G. Robinson, Jr.

The small size of the original breeding nucleus for most research animals raises serious questions about the representativeness of these animals to those in the wild. This is certainly difficult to determine because it is not clear (a) how much inbreeding has been practiced at the Central Laboratory for Experimental Animals or at the West Foundation, (b) what the total population number is and the number of generations involved, (c) how much heterozygosity existed in the original United States group, and (d) how much selection has been practiced and what the frequency of mutation is. According to our own isoenzyme work with blood esterases and other tissue enzymes, the genetic variability is only approximately 5 to 10 percent. Of thirty-nine loci studied in four males and four females, only two were found to be polymorphic. In a wild population, eighteen to twenty segregating loci would be expected. Additionally, Billingham and Seydian (1963) have demonstrated that skin transplants between animals are not rejected for long periods, implying that there are few segregating histocompatibility genes. These findings suggest that inbreeding and inadvertent selection have been instrumental in reducing the overall genetic variance of the species. There is no behavioral or physiological indication, however, that the animals are in any way abnormal.

Life Statistics, Physiology, and Behavior

The Mongolian gerbil has the typical agouti color of most wild rodents. The dorsal hairs have a gray-black base, a yellow center band, and a black tip. The belly is light in color. No coat-color mutants have appeared. The tail is nearly as long as the body and is tufted on the end. The ears are medium long, and the eyes are large and slightly protruding. The hind legs are slightly elongated and modified for jumping.

Vital statistics for physiological processes and events are summarized in Table 5. Data for this table were compiled from reports by Barfield and Beeman (1968), Harriman (1969a; 1969b), Marston and Chang (1965), Rich (1968), Robinson (1969), Schwentker (1963; 1968), Thiessen (1968), Vick and Banks (1969), and Winkelmann and Getz (1962).

Table 5.
Vital statistics for the Mongolian gerbil

Measure	Quantification (Average)
Life span	Over 4 years
Minimum breeding age	10–12 weeks
Maximum breeding age	20 months
Weaning age	21–25 days
Litter size	4.5
Sex ratio	1.03 at birth, 1.00 at weaning
Number of litters in lifetime	6 or more
Estrous cycle	4–6 days
Female receptivity	10–12 hours after light onset
Length of receptivity	5–24 hours
Ovulation	Spontaneous but can be induced
Ovulation rate	4–9 ova
Time of ovulation	6–10 hours after mating
Prenatal mortality	32 percent
Gestation period	24–26 days
Postpartum mating	86 percent fertile
Vagina opens	40–76 days
Testes descend	35 days
Ears open	5 days
Eyes open	20 days
Incisors erupt	12 days
Rectal temperature	38.2° C
Heart rate	360 per minute
Food consumption	5–8 g per day/100 g body weight
Water consumption	4–7 ml day/100 g body weight

Table 6.
Behavioral development of the Mongolian gerbil

Age (*Days*)	Period	Characteristics
0–6	Perinatal	Weak rooting and body righting apparent.
6–15	Neonatal	Young leave nest and respond to maternal odors and to sound (sometimes with weak foot-thumping).
15–18	Transitional	Eyes open and other behaviors become more evident.
18–24	Social	Nursing ceases and play fighting (boxing) occurs among littermates.

Prenatal and postnatal development is somewhat slower in gerbils than in mice or rats, and the lifespan is somewhat longer. The behavioral development according to DeGhett (1964) is indicated in Table 6. The average body weights of adult males and females are 100 and 85 grams, respectively, while the corresponding birth weights are 3.1 and 2.9 grams.

Litter size varies widely, from 1 to 12, with an average of 4.5. Small litters, especially first litters, are often ignored and occasionally cannibalized. Breeding may continue beyond twenty months, and as many as ten litters have been reported for a single female. This is possible because the female has a postpartum estrus allowing blastula implantation during lactation.

Theoretically, then, a gerbil can have a litter every twenty-four days, although in practice delayed implantation follows postpartum mating, thus extending the gestation period. The male may be left with the female throughout the reproductive cycle, although he may take no active part in caring for the offspring.

Table 7.
A summary of estrous events in the female Mongolian gerbil

Stage and Length of Cycle	Vaginal State	Ovarian State	Uterine Weight	Wheel Activity	Behavioral Receptivity
Proestrus *1.6 days*	Nucleated epithelial cells	Medium to large follicles; few corpora lutea	Low	Medium	No
Estrus *1.4 days*	Cornified epithelial cells	Large follicles with some corpora lutea	High	High	Yes
Diestrus *2.4 days*	Leukocytes	Degenerating corpora lutea	Low	Low	No

The estrous cycle of the mature female is highly unpredictable (Marston and Chang 1965; and our laboratory), although both Barfield and Beeman (1968) and Vick and Banks (1969) report fairly regular vaginal and behavioral cycles of approximately four to six days. These cycles are accompanied by successive dominance of leukocytes, nucleated epithelial cells, and cornified cells, much like the pattern in mice and rats. A summary of estrous events appears in Table 7. Our own data would suggest that ovulation is generally spontaneous but can be induced by vaginal stimulation. The dual mechanism could account for the irregularities of cycle length found by some investigators.

Table 8.
Means and standard deviations for body weight and selected organ weights of two-year-old male and female Mongolian gerbils

Measure	Absolute Weight	
	Male	Female
Body Weight (g)	101.30 ± 16.36	87.30 ± 14.10
Liver (g)	6.26 ± 2.02	3.75 ± 1.08
Spleen (mg)	109.60 ± 71.82	83.70 ± 49.18
Adrenals (mg)	37.28 ± 1.43	32.38 ± 1.22
Seminal vesicles and prostate (mg)	967.00 ± 192.23	
Testes (g)	1.40 ± 0.13	
Ovaries (mg)		83.90 ± 24.75
Uterus (mg)		203.30 ± 54.63
Brain (g)	1.00 ± 0.05	1.01 ± 0.04
Thyroid (mg)	12.30 ± 3.05	10.30 ± 2.45
Pituitary (mg)	3.62 ± 1.06	4.07 ± 0.95

Note: Number of animals for each measure ranged from 10 to 55, with a mean of 43.

Adult organ weights for the two sexes are given in Table 8. Sex differences are not particularly large, except for adrenals. Here the sex difference is apparently reversed from that typically found in other rodent species. Preliminary evidence also indicates that cortisol is the principal adrenal steroid in gerbils, whereas corticosterone predominates in mice and rats (Kakihana, Blum, and Kessler, unpublished). Other estimates of organ weight variations appear in sections relevant to behavioral studies and in a report by Wilber and Gilchrist (1965). Morphological and skeletal features are described by Pav and Magalini (1966).

Weight per 100 g Body Weight
Male   Female

Male	Female
6.04 ± 1.27	4.32 ± 1.0l
108.90 ± 74.21	102.50 ± 80.77
52.00 ± 2.00	53.00 ± 2.00
946.70 ± 129.36	
1.36 ± 0.21	
	98.30 ± 29.24
	233.70 ± 61.24
1.02 ± 0.02	1.18 ± 0.18
11.50 ± 4.43	12.50 ± 2.55
3.60 ± 0.94	4.70 ± 1.31

The gerbil must certainly be one of the easiest rodents to maintain and handle: it needs little water and is clean, nearly odorless, and generally free of endemic diseases. Approximately twenty-five square inches of floor space is needed per animal. In general we suggest that the number of square inches per animal correspond roughly to five times the age in weeks for the animals up to seven weeks old and three times for older animals. Wood shavings in regular mouse or rat cages provide adequate bedding and need to be changed only every 1.5–2 weeks if combined with "Litter Green," an alfalfa deodorant of considerable merit and without estrogenic activity (McFadden Industries, Inc., Kansas City, Mo.).

Laboratory maintenance is simplified because of the extensive physiological buffering evident in the gerbil. It can be housed at temperatures up to 97°F for many hours a day for over eight weeks without any apparent difficulty, although, without adequate shade, it will succumb at this temperature. The gerbil is a natural root and seed eater, but it can survive on a wide variety of foods and will select an adequate diet if provided with a self-selection cafeteria (Harriman 1969a; 1969b). Purina mouse and rat pellets provide an adequate diet. Sunflower seeds are a preferred delicacy. Although a gerbil will consume up to six or seven milliliters of water a day, its body weight will stabilize under complete water deprivation at about 11 percent below the normal value (Winkelman and Getz 1962). The animals can derive most of the water they require from the metabolism of foods, especially fats; however, it is our experience that the initial fluid deprivation can be strenuous and should be avoided. High fat diets should be avoided with breeding females, as fat pads accumulate around the ovaries, precluding ovulation.

Aggression is generally low in stable social groups; hence several gerbils can be housed per cage. Eisenberg (1967) refers to the Mongolian gerbil as a "contact" seeker. Isolation tends to increase aggressivity, as does territorial identification, which in our estimation explains the notion that gerbils accept only one mate during a lifetime. Often when a mate dies or is removed, there is an interval before another mate is provided. Separation for only a short time or the identification with a territory may increase aggression and disrupt reproduction. The female seems especially quarrelsome. Mating can proceed without difficulty, however, if isolation is not protracted and if pairs are introduced into strange cages simultaneously. Lifetime monogamy is an artifact of housing conditions.

Gerbils can be handled easily by lifting them near the base of the tail. Holding the tail at the tip may cause the furred tail sheath to slip off. The Mongolian gerbil is less thigmotactic than most rodents; hence it is easy to recover from an apparatus or the floor. It displays little fear in the laboratory and can be enticed within reach by tapping the fingers in its vicinity.

Almost nothing is known about the normal ecology of this species, except that it is a burrowing form which lives in arid regions of northeast China. Seminatural living situations described by Eisenberg (1967) and laboratory investigations reflect only some of its behaviors.

The docility of the gerbil should be a primary factor in increasing its use in laboratory studies. In general, it seems to be a good learner, especially when aversive reinforcers are used. Visual studies indicate acute depth perception and exceptionally low maximum auditory sensitivity (between 3,000 and 5,000 Hz). Ultrasonic emissions are frequent and are now being studied in this laboratory. The gerbil is highly defensive of territories, as this monograph will establish. It therefore combines features valuable for a general laboratory rodent with those highly interesting from a species-common perspective.

Behavioral studies are becoming more numerous. An annotated bibliography of these is presented in Table 9. No attempt is made in this listing to reconcile differences in results or interpretations. Only additional investigations will give clarification.

Table 9.

Summary of behavioral studies with the Mongolian gerbil (*Meriones unguiculatus*)

Type of Behavior	Reference

General characteristics

In their natural environment, the dry areas of Mongolia and northeast China, gerbils build elaborate burrows and are gregarious and active both day and night. The burrows generally have multiple entrances.

Allen (1940); Bannikov (1954); Kasuga (see Schwentker 1963; 1968); Tanimoto (1943); Thiessen (1973)

Gerbils are characterized as "contact seekers." Sandbathing, perineal drag, and ventral marking are periodic and probably related because of natural selection.

Eisenberg (1967)

Gerbils are considered hibernators by some but not by others.

Jettmar (1930); Kalabukhov (1964)

General behavioral and physiological characteristics of the gerbil are summarized and several research advantages are noted.

Nauman (1963); Pav and Magalini (1966); Rich (1968); Robinson (1968; 1969); Thiessen (1968)

Growth, longevity and reproductive life are studied in male and female gerbils.

Arrington, Beaty, and Kelley (1973)

Sensory-motor characteristics

Spectrum analyses on adult gerbils reveal that vocalizations elicited by tail pinching show peaks most frequently between 4,000 and 6,000 Hz, the region in which the cochleogram shows greatest sensitivity.

Finck and Goehl (1968)

Cochlear microphonic becomes evident at 12 days of age in male and female gerbils. Action potentials from the eighth nerve are evident at 14 days, body reflexes to sound are evident at 14 days, and pinna responses appear at 17 days. Maximum sensitivity ranges from 15,000 Hz at day 16 to 4,000 Hz in the adult.

Finck, Schneck, and Hartman (1972)

Cochlear nucleus firing rates are most frequent at tone onset and subsequently decay to a steady state.	Smith and Zwislocki (1971)
Pinnae in male gerbils respond reflexively to a 1,500 Hz tone. The reflex increases in intensity between 50 and approximately 90 dB.	Lippman and Galosy (1969)
Pinnae responses to onset of tones from 500 to 10,000 Hz and from 51 to 101 dB in intensity show a direct relation to intensity but not frequency in males. The responses reflect startle behavior and not auditory sensitivity.	Galosy and Lippman (1970)
Cochlear potential was used to establish greatest auditory sensitivity at 4,000 Hz. The range of sensitivity is 200–32,000 Hz. The unusually low peak sensitivity is associated with an enlarged bulla.	Finck and Soufouglu (1966)
Ultrasounds are produced by adult male and female gerbils between 20 and 150 kHz, with peak intensities up to 106 dB and with a duration from 2 msec to 3.4 sec. Young also emit ultrasounds. A lactating female emitted these signals while thumping with one hind foot. Generally, initial sounds consist of two to five harmonics. Many pulses are within the human hearing range and sound like high pitched "creaks."	Sewell (1970)
Ultrasonic vocalization in gerbils is high from birth until day 4; then the rate declines rapidly until day 10. A moderate decline occurs thereafter and no vocalizations are detected on day 20 or later.	DeGhett (1974)
Male gerbils with bilateral hippocampal lesions are more reactive to auditory and especially visual stimulation. Neocortical damage has no effect.	Ireland and Isaacson (1968)
Hippocampal lesions in male gerbils increase locomotion, rearing, sniffing, and drinking, and decrease sleeping, reclining, and shredding of nest material. Contacts with novel objects are also decreased. Territorial marking is apparently unaffected.	Glickman, Higgins, and Isaacson (1970)

Male and female gerbils show rhythmic slow activity (theta) in the hippocampus during large body movements, such as walking or ventral marking, and large amplitude irregular activity during grooming, chewing, and drinking. Suppressed amplitude activity was observed during concentrated sniffing.	Whishaw (1972)
Both brain-damaged (cortical or septal lesions) and normal males alternate more frequently than they place respond in a cross maze.	Schnurr (1971)
Male gerbils are highly mobile in an open field and spend considerably less time "wall-hugging" than rats. Interaction with conspecifics in an open field decreases with time.	Gerritz (1968); Nauman (1968)
Open-field activity in male gerbils increases with increases in field size from 9 × 18 in to 36 × 72 in and decreases over a 10-min test period.	Oldham and Morlock (1970)
Wire clinging behavior is much less pronounced in gerbils than in mice (*Mus musculus*) and striped hamsters (*Cricetulus barabensis*).	Lerwill (1974)
Young gerbils show increased total locomotion, approaches to stimuli, and total investigative time when the spatial arrangement of objects in an open field is changed.	Wilz and Bolton (1971)
Male gerbils are more exploratory than Sprague-Dawley rats in a Greek cross maze with black, white, striped, and checkered arms. Gerbils prefer the checkered arm and rats the black arm. Handling increases activity for both species.	Thompson and Lippman (1972)
Both male gerbils and Golden hamsters (*Mesocricetus auratus*) show higher cage activity during the dark phase of a 12L-12D schedule, with hamsters showing the more pronounced L-D difference. Feeding behavior shows similar variations.	Lerwill (1974)
In male gerbils the total amount of spontaneous activity in the home cage between 3 and 6 P.M. is in part a function of the change in horizontal magnetic intensity which occurs during this period. The greater the change the more active the gerbil will be, and vice versa.	Stutz (1971)

In contrast to C57BL/6J mice, male and female gerbils show some deficiency in visual cliff behavior. They are highly sensitive to light, however, as they show a strong preference for darkness, and their circadian activity cycle can be readily entrained.	Thiessen et al. (1968)
Male and female gerbils use visual and tactual cues in marking a visual cliff response. Under some situations tactual cues override conflicting visual cues.	Collins, Lindzey, and Thiessen (1969)
Older gerbils fail to alternate spontaneously in a two-choice maze.	Greenberg (1973)
Male gerbils from 170 to 200 days of age show spontaneous alternation in a T-maze. Animals are responsive to their own odor trails but not to spatial cues.	Dember and Kleinman (1973)
Male and female gerbils show an increased latency to jump into an open field as height of the platform increases. They can discriminate alley length (depth) differences of 11.5 and 23 in. with 98 percent accuracy and differences of 15 and 20 in. with 75 percent accuracy.	Cole and Topping (1969)
Early experience with "flat" or "cliff" environments modifies subsequent visual cliff behavior of male gerbils, increasing and decreasing cliff descents, respectively.	Thiessen, Lindzey, and Collins (1969)
When a male gerbil is allowed to regulate its daily light and sound schedule it shows a high preference for light over sound and prefers nearly 23 hrs light per day.	Thor (1972)
Male gerbils foot-stomp following foot shocks, sex behavior, and self-stimulation for electrical impulses.	Routtenberg and Kramis (1967)
Male and female gerbils foot-stomp after posterior hypothalamic rewarding brain stimulation. Stimulation produces hippocampal synchronization (theta); desynchronization occurs during foot-stomping.	Kramis and Routtenberg (1969)

45

A proportion of male and female gerbils seize when exposed to a novel stimulus. Adaptation follows continual exposure. Peak seizure time corresponds to nocturnal activity. Seizures can be counteracted with diphenylhydantin (7.5 mg/animal).

Thiessen, Lindzey, and Friend (1968)

Spontaneous seizures in gerbils are counteracted by trimethadione (300 mg/kg) or phenobarbital (15 mg/kg), but not by diphenylhydantin (75 mg/kg).

Goldblatt et al. (1971)

Spontaneous seizures in gerbils appear at 2 months of age and increase in frequency to at least 6 months. Early tests delay the appearance of later seizures. No sex differences are apparent.

Kaplan and Miezejeski (1972)

In gerbils selected for seizure susceptibility, seizures start at 9 weeks and reach 97 percent at 6 months. Severity increases during the same period. Death because of seizures is rare.

Loskota, Lomax, and Rich (1974)

Seizures in gerbils are not reliably triggered by electrical-shock, photic, or auditory stimuli. Shocks may abort seizures in progress.

Goldblatt (1968)

Seizure-susceptible and nonsusceptible gerbils were selected for over a 3-year period. The rates of seizure are 90 percent and 8 percent, respectively. No sex differences are evident. Repeated testing results in increased latency of onset but in longer seizure duration.

Loskota, Lomax, and Rich (1972)

Male and female gerbils shred paper for nesting material. Preference for the paper increases as it increases in thickness from .002 to .045 in.

Glickman, Fried, and Morrison (1967)

Body temperature of gerbils 2 to 25 days of age ranges from 27.8 to 33.0° C in the nest. This is below the adult temperature of 38.2–38.6° C. When the young are exposed from 1–2 hrs to 14–15° C, gerbils 12 days old or younger are unable to thermoregulate. Between 13 and 20 days there is a linear increase in endothermic capability which reaches its peak of 35° C at 21 days. This is still below adult levels.

McManus (1971)

The resistance to cold was studied in 8 rodent species including the gerbil. The greatest resistance to −40° C was found in the collared lemming.

Ferguson and Folk (1970)

Gerbils, like mice, show a developmental change in temperature preference (36–38° C) from birth to maturity. This appears to be correlated with stabilization of metabolic rates and may also be associated with growing independence of maternal heat.

Eedy and Ogilvie (1970)

The active fraction in the sebum of the ventral scent gland is phenylacetic acid. Both the fraction and a synthetic sample elicit discrimination in a Skinner box and active exploration.

Thiessen et al. (1974)

Learning abilities

General aspects of elicited behavior and reinforcement are discussed.

Glickman and Higgins (1968)

Male and female gerbils readily learn two- and four-choice object discriminations on WGTA (Wisconsin General Test Apparatus). Fewer errors are committed on the two-choice problems.

King, Goodman, and Reese (1968)

Male gerbils form learning sets on a WGTA at a somewhat slower rate than cats or marmosets.

Blass (1965); Blass and Rollin (1969)

Male gerbils acquire a light-cued discrimination response to food in a Y-maze at a somewhat slower rate than Sprague-Dawley rats.

Wise and Parker (1968)

Male gerbils learn a passive-avoidance shock response at the same rate as Swiss Charles River mice and golden Syrian hamsters, but at a slower rate than "hooded" or Holtzman albino rats.

Walters and Abel (1971*b*)

Rats outperform gerbils in the initial trials of a step-through passive-avoidance task. Habituation to handling facilitates learning in both species.

Galosy, Lippman, and Thompson (1970)

A tone followed by shocks inhibits sand-digging but increases "alert posturing" in male gerbils. Once conditioning is established the onset of tone is avoided by lever pressing. Foot-stomping follows shock. Neither sand-digging nor posturing fully returns to base levels during extinction.	Walters and Glazer (1971)
Pyrahexyl (2.3 mg/kg body wt), a synthetic cannabis, slightly facilitates shock-avoidance learning in male gerbils.	Walters and Abel (1971*a*)
Gerbils can discriminate between synthetic cannabis and pentobarbital in a shock apparatus.	Jarbe, Johansson, and Henriksson (1975)
Male and female gerbils, kangaroo rats (*Dipodymus merrami*), and mice (*Mus musculus*) were compared for shock avoidance in a two-way shuttle box. Order of acquisition was mice, gerbils, and kangaroo rats.	Boice, Boice, and Dunham (1968)
Male and female gerbils outperform rats in a one-way shuttle box apparently because of fewer displacement responses.	Ashe and McCain (1972)
Shock-avoidance learning emerges only in groups of gerbils exposed to an escapable or a brief inescapable shock. Response termination of US (unconditioned stimulus) is not necessary for shuttle-avoidance learning.	Galvani (1974)
Male and female gerbils respond to 24-hr food and water deprivation by pressing a button to obtain chocolate milk. A fixed ratio (FR of 1 response), a 15-sec inter-trial interval (VI = 15), and a 1-min variable interval (VI = 1) were used. Response ratio from high to low is VI = 15, VI = 1, and FR.	Campbell, Straney, and Neuringer (1969)
Gerbils and rats compared in an operant reinforcement situation do equally well for water and food reinforcement on either fixed-ratio or variable-ratio reinforcement.	Vander Weele, Abelson, and Tellish (1973)

Adult gerbils (sex unspecified) acquire bar-press-ing for water or food pellets under a variety of schedules of reinforcement (FI [fixed interval], VI [variable interval], DLR [delayed response], and mixed schedule). Ten-to-20–min experiment-al sessions maintain responding at reliable levels for months.

Vander Weele and Abelson (1973)

Male and female gerbils are generally superior to Sprague-Dawley albino rats in acquisition of Sid-man free-operant avoidance. Performance is best when the shock interval is shorter than the re-sponse-shock interval.

Powell (1971)

The effects of food deprivation and drive shifts were studied on the acquisition and extinction of male gerbil runway performance. Twenty-two–hr deprivation is superior to 6- or 12-hr.

Reynierse, Scavio, and Ulness (1969)

Gerbils and kangaroo rats (*Dipodymus merrami*) make more errors in a complex maze when their eyelids are stitched shut or when their vibrissae are removed. The combined operation results in the greatest number of errors.

Webster and Caccavale (1966)

Male gerbils receiving shock-escape training re-tain the response rather than abandon it when punished.

Martin, Ragland, and Melvin (1970)

Male gerbils acquire a step-through passive-avoidance task at a slower but more constant rate than male Sprague-Dawley rats. Habituation to handling facilitates acquisition in both species.

Lippman, Galosy, and Thompson (1970)

Avoidance behavior of male gerbils in a level press shock box is superior to that of rats.

Walters, Pearl, and Rodgers (1968)

Male gerbils show a positive relationship between shock-avoidance learning and percent reinforce-ment (100, 50, or 0 percent). Those receiving par-tial reinforcement are most resistant to extinction. More rapid extinction is obtained under CS-US (conditioned stimulus–unconditioned stimulus) pairing than regular avoidance extinction.

Galvani (1971)

Male and female gerbils are more active in wheels than rats but learn unsignaled (Sidman) avoidance more slowly. No warm-up effect is apparent, and extinction is slow. Mice acquire a two-way shuttle avoidance more quickly than gerbils.	Powell and Peck (1969)
Male and female gerbils were compared to albino, hooded, and black rats on a Sidman shock-avoidance task. Neither the gerbils nor the black rats (semidomesticated species) demonstrated warm-up effects, as compared to the albino and hooded rats (domesticated species). The gerbils generally attained criterion performance most rapidly.	Powell (1972)
Male gerbils differentially lay odor cues in a runway under reinforced and nonreinforced conditions. Rats can follow these cues to maximize their chance of reward.	Davis et al. (1970)
Male and female gerbils respond to odor cues left by gerbils that were either reinforced (R) or nonreinforced (NR) at the end of the alley. Running time is greater under NR conditions.	Topping and Cole (1969)
Male and female gerbils demonstrate conditioned suppression of a running response.	Frey, Eng, and Gavin (1972)
Gerbils (sex unspecified) given septal lesions are deficient in a passive-avoidance task and perseverate in running to a drinking tube.	G. E. Brown (1972)

Consummatory responses

Gerbils consume water when it is available (0.039 g/g body weight/day), lose weight without it, but stabilize after 29 days. Salt solutions up to 0.8 molar can be utilized.	Winkelman and Getz (1962)
Male and female gerbils and kangaroon rats (*Dipodymus ordii*) consume 0.09 and 0.11 cc water/g body weight/day respectively. When water was withheld, sunflower seeds retarded body-weight loss, but the animals were dying by the time their weights were down about 30 percent.	Boice and Arledge (1968)

Male and female gerbils respond to 13 days of complete water deprivation with 17 percent loss in body weight. Body weight is maintained by 2 ml water/100 g weight daily.

Arrington and Ammerman (1969)

Adrenalectomized male and female gerbils failed to regulate sodium dysfunction by a compensatory increase in NaCl (0.15M or 5 percent). Death can be prevented by injecting hydrocortisone acetate deoxycorticosterone.

Cullen and Scarborough (1970)

Male gerbils, like rats, respond to formalin injections with an increase in NaCl drinking. Gerbils differ from rats in that the increase takes longer and is not dose-dependent. Control animals injected with isotonic saline also show increased NaCl ingestion, suggesting a different homeostatic mechanism in desert mammals.

Cullen (1972)

Male and female gerbils show a salt-solution preference of 0.9 percent even though they can utilize salt solutions as high as 6 percent.

Grimsley (1973)

In 4 species of gerbils (*Meriones unguiculatus, M. libycus, M. shawi*, and *Psammomys obesus*) sucrose is one of the most effective stimulants for drinking. The smallest NaCl response occurs in gerbils that are subject to the greatest osmotic stress in their natural environment, relative to rats.

Jakinovich and Oakley (1975)

Male and female gerbils offered a selection cafeteria of casein, sucrose, salt, and cottonseed oil, along with vitamins in water, select an adequate diet. The diet results in weight gain over Purina Chow.

Harriman (1969 *a, b*)

Gerbils (sex unspecified) lose more body weight under food-deprivation conditions than under water-deprivation conditions.

Dunstone, Krupski, and Weiss (1971)

Male and female gerbils were placed on restricted feeding (0, 40, 80, and 100 percent of usual daily consumption). Water intake shows a short period of adipsia followed by polydipsia under 0 and 40 percent food deprivation. The polydipsia is permanent.

Vander Weele and Tellish (1971)

Male gerbils respond to 2–5 days of food deprivation with polydipsia (35+ ml/day), which returns to normal (10 ml/day) when food is restored. Polydipsia is not related to Na^+ deficiency, and urine output is hypertonic when salt is injected. Hamsters, but not guinea pigs and rats, are known to respond similarly.

Kutscher (1969); Kutscher, Stillman, and Weiss (1968)

Female gerbils hoard more food pellets than males. Male castration increases hoarding to female level and testosterone replacement lowers levels.

Nyby et al. (1973)

D-amphetamine is more potent than the l-form in suppressing feeding and food hoarding (12 mg/kg).

Nyby, Belknap, and Thiessen (1974)

Intraperitoneal injections of atropine sulfate or atropine methyl nitrate decrease water consumption in male and female gerbils, suggesting cholinergic mediation of drinking.

Milner (1972)

Intracranial injections of angiotensin II elicit polydipsia in normal or excessive lettuce-eating in water-deprived gerbils. Norepinephrine is also dipsogenic, while carbachol is without effect on ingestive responses.

Block, Vallier, and Glickman (1974)

Reproductive patterns

The basic mating pattern of gerbils is similar to that of other mammals, with a refractory period following successive ejaculations. Number of intromissions to ejaculation is extremely high. Foot-stomping is higher in males than in females, increasing after ejaculation and decreasing with successive ejaculation.

Kuehn and Zucker (1966; 1968)

The estrous cycle of female gerbils is determined to be 4–6 days, using indicator male. Estrus, lasting about 12–15 hrs, is generally preceded by small cornified vaginal cells called "transitional" cells.

M. A. Barfield and Beeman (1968)

The estrous cycle of female gerbils is determined to be 5.34 days, using vaginal state, wheel revolutions, and receptivity to males. Uterine weight doubles at estrus, and ovarian follicles reach maximum size shortly before onset of estrus.

Vick and Banks (1969)

The interval between pairing and mating in male-female pairs is shortest (40 days) when mature, experienced males are involved. Thirty- but not 15-day separation between pair members leads to fighting.

Norris and Adams (1972)

All females show postpartum estrus. Corpora lutea are prolonged by the presence of the litter (26 days) or in the presence of a male (13.5, vs. 8.8 days when male is absent). In the latter condition an intact male (vs. vasectomized or no male) hastens second postpartum estrus.

Meckley and Ginther (1972)

Excision of the ventral sebaceous gland in male and female gerbils does not preclude successful reproduction.

Mitchell (1967)

Approximately 80 percent of female gerbils mate postpartum: 59 percent maintain pregnancy to term when the first litter remains, as compared to 80 percent when the young are removed at birth. Removal of litter on days 5, 10, or 21 from a female nursing 4 pups results in gestations of 25, 30.2, and 36.9 days, respectively. Implantation usually occurs 8 days after mating but can be extended to day 25 in lactating females. FSH (125 μg-250 μg) facilitates implantation and is evident in all females by day 6 postpartum.

Norris and Adams (1971)

Water deprivation produces deficits in reproduction in fertile pairs, but reproduction will continue if only the male is deprived. In the latter case ventral marking and secondary sex characteristics are depleted: all effects are reversed by reinstating water balance.

Yahr and Kessler (1975)

Social behavior

Gerbils show a progressive increase in complex behavior from the perinatal period (0–6 days) and neonatal period (6–15 days) to the socialization period (15–18 days). Their eyes open between days 15 and 18, and play fighting begins on day 19.	De Ghett (1964)
Gerbil pup development advances sequentially, with infants spending most of their time sleeping or nursing until day 18, when many new patterns of behavior emerge, including grooming, increased activity, and wrestling. Handled pups spend more time quietly.	Kaplan and Hyland (1972)
Rat-reared gerbils show early weight gains and less locomotion and rearing in open field. Ventral marking is not affected.	Bols and Wong (1973)
Unlike rats, male gerbils persist in fighting behavior between shock intervals in a paired shock-induced fighting situation.	Dunstone et al. (1972)
Pairs of shocked gerbils show agonistic behaviors until dominance order is established. Males win encounters over females as a reliable dominance hierarchy emerges.	Boice and Pickering (1973)
Gerbils, like mice and especially rats, respond to enriched living conditions with increases in cortical weight, ratio of cortical/subcortical weight, and ratio of cholinesterase/acetylcholinesterase in cortex. Cortical acetylcholinesterase per unit weight is decreased.	Rosenzweig and Bennett (1969)
Gerbils choose their home cage rather than a strange cage on initial preference tests. This response habituates over 6 daily sessions but can be reinstated following mild foot shcok.	Ginsburg and Braud (1970)

Male and female gerbils discriminate between individual gerbils and their urine in a two-choice olfactory-discrimination task. The urines may be diluted as much as 600–800 to 1 and discrimination continues. When the discrimination is between dilute urine and nothing, dilution is as much as 7,000 to 1. Gerbils also readily discriminate between other gerbils and small rodents of other species.

Dagg and Windsor (1971)

Adult males can discriminate between conspecifics, using ventral gland secretions, urine, or soiled shavings. No discrimination is evident for fecal pellets.

Halpin (1974)

Gerbils maintained in various numbers in the same cage (13 × 15 × 8 in.) exhibit little aggression, and territorial responses appear to be absent.

Fisler (1970)

Isolation of gerbils from 15 to 170 days of age depresses ventral marking, aggression, and open-field activity. All isolates display spontaneous seizures.

Berg, Shanin, and Hull (1975)

In an empty room, male gerbils establish a dominance order but do not defend territories. Dominant males are the only ones observed to mark; however, they neither mate nor interfere with mating by subordinates. Males foot-stomp more than females, especially during sexual bouts.

Gallup and Waite (1970)

Male and female pairs exclude either sex from their territory. Migrants will swim a water barrier to escape territorial holders and will avoid olfactory cues from that territory.

Thiessen and Dawber (1972)

Male intruders subject to attack by residents decrease scent marking in the foreign territory and in areas visually resembling that territory. Olfactory cues are not involved.

Yahr (in press a)

Adult male gerbils show individual differences in competition for various incentives. The differences seem to depend on skill rather than dominance relation.

Wechkin and Reid (1970)

Territorial marking and ventral scent gland size are sexually dimorphic; the male marks about twice as often as the female and has a gland roughly twice as large. Ventral marking in males is positively correlated with urination and defecation.

Thiessen, Blum, and Lindzey (1969)

Adult female gerbils that mark at relatively low levels do not show a decrease in ventral marking following gonadectomy and/or adrenalectomy. Moreover, in females ovariectomized at 22 days of age, marking develops normally. The ventral gland, however, is smaller in ovariectomized females.

Owen (1972); Whitsett and Thiessen (1972)

Male castrate ventral marking is stimulated by 640 μg of testosterone, which is negated by 3 but not 1 mg of progesterone.

Griffo and Lee (1973)

Estrogen (40 μg/wk) or estrogen plus progesterone (40 μg + 25.0 μg/wk), but not progesterone alone, will stimulate ventral marking in ovariectomized females. Females that have never marked are unresponsive to steroids.

Yahr and Thiessen (1975)

Ovariectomized females respond with ventral scent marking to testosterone or estrogen with testosterone. Androsteredione increases marking in intact females but not in ovariectomized females.

Owen and Thiessen (1973)

Treatment of ovariectomized females with 20, 40, or 80 μg of estrogen results in ventral scent marking for every dose of progesterone administered (0, 250, 500, or 1,000 μg). Adding progesterone to long-term estrogenized ovariectomized females evokes greater marking.

Owen and Thiessen (1973)

Adult nonlactating females do show a significant decrease in marking following gonadectomy, provided that their original levels of marking were high. Various combinations of estradiol benzoate (EB) and progesterone (P) reinstate the behavior.

Owen (1972)

Small doses of TP injected into the lateral ventricles or testosterone implanted with cannulae in
the hypothalamus can evoke marking behavior in
castrated males. Actinomycin D, when combined
with TP in the ventricles, blocks the induction of
marking.

Thiessen and Yahr (1970)

In intact females, TP injections elevate marking
above normal male and female levels and increase gland size. Regardless of hormone state
there is a strong relation between marking, gland
size, and sebum secretion (gland blot).

Thiessen and Lindzey (1970)

Female gerbil scent marking, gland size, and nest
building increase during gestation and lactation.
Ovariectomy attenuates the increases in marking
and gland size. Lactating females retrieve pups
marked with their own sebum before retrieving
others.

Wallace, Owen, and Thiessen (1973)

Implants of estrogen or testosterone into anterior
hypothalamus of ovariectomized females stimulates ventral marking. Estrogen is most effective
in anterior hypothalamus, preoptic area, and
septum. Similar implants into thalamus, hippocampus, amygdala, and anterior olfactory nucleus are ineffective.

Owen, Wallace, and Thiessen (1974)

Eleven steroids were tested for effects on ventral marking of castrate males (systemic injections or preoptic implants). Testosterone is the
only endogenous steroid to elicit marking, but related steroids (with a 17 β-hydroxyl group and a
double bond involving carbon 4) are also effective.

Yahr and Thiessen (1972)

Male castrate ventral marking is reinstated by
testosterone or estrogen but not dihydrotestosterone. The nonaromatizable steroids 6 α-fluorotesto-
sterone and 1 α-methyltestosterone stimulate
marking or have no effect, respectively.

Yahr (1976)

Males from which the ventral gland is excised
show the same frequency and temporal pattern
of marking as sham-operated and skin-excised
males. Prior marking experience in the apparatus does not affect the results.

Blum and Thiessen (1970)

Territorial behavior in two females gerbils is
studied.

Rothenberg (1974)

Male gerbils and hamsters (*Mesocricetus auratus*) were tested in groups within species, in tube
competition for food and in open competition for
water. Dominance readily formed in both species.
but was not evident for two situations in the gerbil;
it was positively related in the hamster. Body
weight in part determined water dominance.

Boice, Hughes, and Cobb (1969)

Fighting is higher in male-male (70 percent) than
in male-female (16 percent) or female-female (10
percent) pairings. Anesthesia before grouping reduces aggression in pairs (83 percent) but not in
groups of 4.

Norris and Adams (1972)

Male gerbils establish linear dominance orders in
competition for food or for access to an un-
charged platform suspended over a shock grid.
There is moderate between-individual reliability
within problems but not across problems.

Wechkin (1964)

Adult or infant male castration leads to an increase in aggression which is reversed by testo-
sterohe replacement. Testosterone inhibits aggression in females.

Anisko, Christenson, and Buehler
(1973)

Weekly injections of 640 μg of testosterone propionate (TP) increase marking frequency in intact
male and female gerbils when objects are close
to the floor. Gland size also increases with TP administration. Gerbils chin higher objects with a
sebaceous gland complex located on the chin
and neck.

Thiessen, Yahr, and Lindzey (1971)

Male gerbils are highly exploratory and engage in
various agonistic behaviors. Male-male encounters are characterized by ambivalence followed
by aggression and finally dominance relations.

Reynierse (1971)

Marking behavior and scent gland integrity are under androgen control, as indicated by castration and TP replacement (80 or 640 μg). Scent gland integrity seems less important for marking than hormone influence on the CNS (central nervous system), although the gland can be used as an external measure of androgen levels.

Thiessen, Friend, and Lindzey (1968)

Marking behavior and the scent gland are entirely absent in young male castrates. Relative to controls, hormone-treated (640 μg TP) males begin marking earlier and reach higher frequencies. Gland development is also responsive to TP but lags behind marking activity.

Lindzey, Thiessen, and Tucker (1968)

Male and female gerbils begin to mark and build cotton nests at 5 weeks of age. Sex differences in marking appear at 16 weeks, with males marking more; no differences appear in nest building.

Lee and Estep (1971)

Male gerbils are attracted to the sebum from other gerbils, especially males, but not to their own. Ventral gland excision reduces territorial marking somewhat. Olfactory bulbectomy eliminates marking in 93 percent of tests. Marking is lower in the dark.

Baran and Glickman (1970)

Adult male gerbils prefer to investigate areas with self-generated or conspecific odors over those without odors. The preferences are dependent upon intact olfactory bulbs but still occur without visual input. There is evidence that preferences are somewhat dependent upon testicular secretions.

Baran (1973)

Olfactory bulb aspiration almost completely eliminates marking behavior. The reduction is apparently independent of any endocrine dysfunction, as indicated by weights of seminal vesicles, testes, and adrenals. Marking, however, can be at least partially reinstated with massive doses of TP (640–1280 μg). Also, 80 μg EB, but not 800 μg P, significantly depresses marking in males. EB decreases testis and seminal vesicle weights and stops spermatogenesis.

Thiessen, Lindzey, and Nyby (1970)

Males with peripheral anosmia (5 percent zinc sulfate into nasal cavity) show reduced ventral scent marking.	Wallen and Glickman (1974)
Grouping of males increases aggression. Anosmic animals (olfactory bulbectomy or ablation of olfactory receptors) are deficient in social interactions, whereas bulbectomy enhanced aggression.	Hull et al (1974)
Both TP and EB increase scent marking in castrated male gerbils, though TP appears to be more effective than EB. The TP effect is attenuated when the two hormones are administered together, suggesting competition for the same receptor sites.	Nyby and Thiessen (1971)
Hypothalamic implants of testosterone or TP induce marking in castrated male gerbils. TP implants in the hippocampus, amygdala, septum, reticular formation, caudate, or cortex are not effective. Large preoptic-area hormone implants are more effective than small ones (21, 23 vs. 25, 27 gauge). Small amounts of actinomycin D (2.0–2.5 μg/μl) or puromycin 1:3, TP by weight) inhibit marking when combined with the hormone. RNase has a similar, but short-lived, effect. Dihydrotestosterone does not induce marking when implanted in the preoptic area. Magnesium pemoline (1:1, with TP by weight) suppresses TP effects in the preoptic area but enhances marking somewhat by itself.	Thiessen, Yahr, and Owen (1973)
The antiandrogens, cyproterone and cyproterone acetate, do not interfere with male aggression or territorial marking. They do cause regression of the seminal vesicles and marking gland, however. Both behaviors diminish after castration and reappear with TP treatment.	Sayler (1970)
Castration of adult male gerbils increases aggression between males. Blinding results in a slight increase in aggressive interactions and a marked increase in locomotion. Anosmia produced by bulbectomy or with zinc sulfate results in a general reduction in social interaction and a moderate increase in locomotion.	Christenson et al. (1973)

Gerbils (sex unspecified) respond to 35 days of grouping (N=6) with a fall in adrenal ascorbic acid from 163.9 to 119.1 mg/100 g body weight.

Hughes and Nicholas (1971)

Ventral marking by male gerbils is related to social dominance in pair and triplet encounters. Triplet grouping and population density depress marking and morphological indices of androgen activity. Related studies demonstrate that males are attracted to sebum from the scent gland of other males but not to sebum from females. Females show no preference.

Thiessen et al. (1970)

In same-sex but not mixed-sex groups of gerbils, crowding depresses scent marking, ventral gland and testis weights.

Hull, Langan, and Rosselli (1973)

Males were tested for aggression toward strange conspecifics in a familiar and an unfamiliar territory. Aggression was higher in the familiar environment, suggesting that territorial defense is characteristic of this species.

Wechkin and Cramer (1971)

Males and females were maintained in sensory contact with or sensory isolation from the opposite sex during development (physical contact excluded). Isolation results in more fighting. Males reared with females mark more.

Rieder and Reynierse (1971)

Castration eliminates the scent gland in both male and female gerbils but decreases marking only in the male. Both sexes, on the other hand, are responsive to TP injections. Adrenalectomy, either alone or in combination with ovariectomy, has no effect on marking in the females. Social competiton significantly elevates ventral marking. A low but significant correlation exists between male marking frequency and the assumption of dominance. But, regardless of initial marking frequency, males which become dominant in a competitive situation mark significantly more often than submissive males. In social competition across two territorial boundaries, the male that becomes dominant exhibits a significant preference to mark his opponent's territory before marking his own.

Thiessen, Owen, and Lindzey (1971)

Male gerbils were tested for marking frequency in an occupied territory while the residents were constrained. Following each test, the residents were allowed to attack one group of males (intruders) but not the other (controls). The marking levels of intruders dropped to 25 percent of control levels but returned to normal in novel territories. Olfactory cues from residents or other intruders depressed marking by intruders. Moreover, intruders avoided residents' odors in a Y-tube preference test. Autopsies revealed no evidence of androgen inhibition or stress effects.

Nyby, Thiessen, and Wallace (1970)

Male and female gerbils attack intruder litter mates only if isolation is extensive. Foot-stomping occurs when attacks are not immediate, and submissive postures allay attacks.

Ginsburg and Braud (1971)

The presentation of a gooselike or hawklike figure evokes orientation and fear reactions (including a reduction in marking) from male and female gerbils. Adaptation occurs with repeated trials, especially for the hawklike stimulus.

Bauer (1970)

3

Social Behavior and Evolution of the Gerbil

Naturalistic Observations

Very little is known about the natural characteristics of the Mongolian gerbil. Because it is from the remote areas of eastern Mongolia, northeast China, and western Manchuria, no systematic reports have appeared in the literature. At these latitudes there is an annual rain and snowfall of approximately ten to twenty inches, 75 percent of which occurs during three summer months, beginning with the spring monsoon. Temperature variation is likely to be extreme, but we have no definite records. For the most part, then, the gerbil is known to live in an arid environment with seasonal peaks in rainfall and temperature.

Tanimoto (1943) has briefly reported on the gerbil's burrowing system. According to this investigator the animals live as family groups in elaborate burrows, sometimes reaching three feet in depth. Within the subterranean system there are several spacious areas larger than the tunnels which are used for nesting and food storage. Several openings penetrate to the surface of the ground.

Bannikov (1954) reports that gerbils breed during the spring and summer. Females generally bear two litters each year and the young spend their first winter in the maternal nest with both parents. Gerbils eat green plant food when it is available in spring and summer but preferentially eat seeds the rest of the year. Beginning in August they spend much of their time hoarding food. Also, according to Banni-

kov, many gerbils live in human dwellings. Probably, as with *Mus musculus*, wild and commensal forms are developing. Certainly it is imperative that we learn more about the aboriginal gerbil, and any research in that direction should be encouraged.

Seminatural environments have been developed for the gerbil by a few investigators, but as yet the reports are fragmentary and incomplete. Eisenberg (1967) reports that under conditions of "free living and reproduction" the gerbil is a "contact seeker," often aggregating peacefully and showing small personal distances. Sandbathing, perineal dragging, and ventral marking are periodic and are probably related, since they often occur in temporal proximity. The male displays many mounts and intromissions prior to ejaculation in a copulatory sequence. Laboratory findings (Kuehn and Zucker 1966) and our own observations support Eisenberg's findings, but other investigators report that the mating pattern of the male gerbil is typical of that for cricetid rodents (Davis, Estep, and Dewsbury 1974).

We have attempted to establish gerbil colonies at the Brackenridge Research Station, University of Texas at Austin campus, with the kind permission of Frank Blair of the Department of Zoology. The areas are ideal for field observations, as several walled territories, approximately forty feet square, are available. Animals are restricted to these areas and only rarely egress over a four-foot-high retaining wall.

Some of our attempts have not been productive, but occasionally animals have survived long enough to provide interesting information. Ching-Tse Lee and John Nyby were primarily responsible for data collection. Although the data are not statistical in nature, they are still intriguing enough to summarize.

First, it proved difficult to establish animals in the areas. Predation from cats, owls, hawks, and roadrunners took its toll, and rain prevented survival in some cases. We found that survival was enhanced by digging an initial shelter for the animals and allowing them to complicate it. We also provided food periodically, such as sunflower seeds and rice, although they eventually learned to survive on their own. In one case we covered an area to prevent predation. In all, perhaps three or four pairs had sufficient longevity to allow observations over a period of weeks or months.

The clearest results were these: Gerbils extended their burrows into complex matrices with as many as six entrances. Additional burrows were also dug, such that a single group of gerbils sometimes used five or more burrows. The internal temperatures of the burrows during cold periods were as much as 22°C higher than the ambient temperature. On the surface, paths radiated from the main burrow into the underbrush and along the walled perimeter. The entire area was apparently explored via these paths. Peculiarly, at intersecting trails grass blades were deposited. The paths were generally well-hidden and provided avenues of retreat to the burrow.

Once gerbils were released in an area, they were rarely seen unless the observer waited patiently for long periods. While we suspect that gerbils are primarily nocturnal, they did appear above ground during the day. In general, the docility common in laboratory animals was quickly lost in the field, and they reverted to what is probably their natural wariness.

Foraging for food was common. Animals approached a source, such as sunflower seeds, placed the food in a cheek pouch, and scurried back to the burrow. They seem to be natural hoarders. They ingeniously found food sources and overcame unusual obstacles to reach them. Movies were made of gerbils regularly climbing a tree in order to scale a two-foot fence to obtain sunflower seeds. Their arboreal agility is astounding.

We have little information on the usual size of the family unit. Only one pair gave birth in the outdoor regions and raised their pups to maturity. It looked as though the young migrated at weaning to a different area of the enclosure and started a new burrow. We cannot be certain, however, as a heavy rain killed the animals before they could become established.

Territorial defense against intruders was clear-cut after a family unit was established. Strangers of both sexes were viciously attacked when introduced into the living area. Interestingly, the strangers were not only attacked when near the burrow system but were chased for long distances. The pursuit was savage and continuous and did not depend on the sex of the intruder. Of the several animals introduced, it was obvious that none would have survived to establish residency in the occupied territory.

We have been continuing similar observations under seminatural conditions in the laboratory. Four social-naïve pairs were introduced into a U-shaped aquarium measuring approximately 10 feet in length, eight feet in width, and four feet in depth. Two feet of sandy loam covered the floor of the aquarium. Within one week after introduction one social pair was formed, and this pair henceforth controlled the entire area. This pair dug elaborate burrows and fought off all attempts of other males or females to establish themselves. The dominant pair hoarded all food pellets placed on the surface and actively prevented subordinate animals from eating. The latter animals were also prevented from burrowing, grooming, or scent marking. For the most part the subordinate animals were relegated to one corner of the environment and forced to remain inactive. Within three weeks all subordinate animals were killed outright or starved to death.

Within six weeks after pairing the dominant animals gave birth to a litter of four. The young stayed under ground until about twenty-five days of age. Currently the original pair and the four adolescent offspring are living harmoniously. Seventeen burrow entrances exist, with all burrows interconnected. The burrows run at various depths, some extending the length of the aquarium. Many are exposed by the sides of the glass. The glass has been obviously smeared heavily by body exudates.

Observations and video tape recording indicate that all behaviors we have studied in the laboratory are apparent in this situation. The animals are mostly nocturnal, although they can be enticed to the surface during the day by throwing food pellets into the aquarium or by making noise. Observations are continuing.

Among gerbils living on the laboratory floor a similar social organization is apparent. Of the several animals observed, some have assumed definable territories and defended them with great vigor. In some cases we have noted that a male and female will pair up and live together. We have one interesting case in which a male and female took up residence in one room approximately ten feet square and actually controlled two adjacent rooms of similar size and the connecting hallways. All gerbil intruders, and in some cases humans as well, were attacked. Gerbil intruders were chased by the male for long distances into a connecting colony area. The male often patrolled the territory, initiated the attacks, and did the chasing. The female, on the other hand, spent a large share of her time hoarding food pellets from colony cages into a closet which the pair used as a nesting area. The stash of food grew to over fifty pounds within a few weeks. As far as we could tell, almost all of the hoarding was done by the female.

Hoarding is a predominant behavior of the female gerbil. On several occasions in the laboratory, gerbils have accumulated large stocks of food in closets and beneath and behind cage racks. In controlled experiments, it is apparent that the female normally performs most of the hoarding, while the male will occasionally engage in the behavior (Nyby, Wallace, Owen, and Thiessen 1973), although in the seminatural environment both sexes hoard. These experiments suggest that the relative lack of hoarding in males is androgen-dependent, since castration substantially increases the behavior. A recent study indicates that both *d*- and *l*-amphetamine will suppress hoarding and feeding, with the dextro form being more potent (Nyby, Belknap, and Thiessen 1974). Nyby is continuing his investigation of the possible hormone influence in both the male and the female.

Overall, our seminaturalistic observations have been most rewarding. It is clear that gerbils do assume and defend territories and that there is sometimes a sexual differentiation of social responsibilities. Males are evidently the more aggressive, although the female occasionally participates in territorial defense. For the most part, however, the female's activities are restricted to hoarding, while the male patrols the territory. Pair-bonding occurs, although we are not yet convinced that gerbils are completely monogamous: they may normally live as family units. The major conclusions reached by informally observing their behavior in the Brackenridge area and in the aquarium and on the laboratory floor have been confirmed and amplified by experimental work.

Laboratory Investigations

Characteristics of the Mongolian gerbil described above have posed numerous experimental questions that can only be answered with laboratory techniques. Laboratory studies of behavior, we realize, are not definitive and must eventually be extended by observing animals in their natural environment. They can, however, outline the essential parameters of behaviors and suggest how social behaviors are organized at the physiological level.

We have attempted to structure our laboratory studies in such a way that they capitalize on the species-specific behaviors of the gerbil and at the same time allow us to ask very specific questions about the mechanisms underlying territorial behavior. This and the two following chapters outline our findings on the social consequences of territorial scent marking. We later speculate on the evolutionary implications of the behavior (Ch. 7). In total the studies touch upon many aspects of importance in understanding territorial behavior and scent marking. They also complement the naturalistic observations that we have on hand and outline quite clearly what we additionally must know about the gerbil's natural behaviors.

We have been fortunate to find that male and female gerbils frequently scent mark almost any object in their environment with a ventral sebaceous gland. They mark their pups, each other, and almost any protruding object available. Hence, it has been possible to measure scent marking in a standardized apparatus that permits the animals to direct their marking toward specified objects during a set unit of time. Marking consists of lowering the ventral gland onto an object as the gerbil moves forward over it. The behavior is discrete and easily tallied.

A

Figure 2.
Major phases of marking: approach (*A*) and marking (*B*).

Unless otherwise noted, we typically record ventral marking frequency during five-minute tests in an open field studded with small pegs. The apparatus and behavior are shown in Figure 2. The apparatus is a grey wooden open field one meter square which is marked off into sixteen squares of equal size. A 0.16-centimeter-thick sheet of clear Plexiglas covers the floor, and a roughened Plexiglas peg 2.6 centimeters long, 1.2 centimeters wide, and 0.7 centimeters high is positioned at each of the nine line intersections. A fifteen-watt fluorescent tube, positioned 1–1.5 meters above the floor of the apparatus, provides the only illumination in the testing room.

B

Ventral marking is directed almost entirely at the Plexiglas pegs. We record general
activity by counting the number of lines
crossed during a set period of time. Often
urine pools and fecal boluses deposited
during the trials afford additional measures
of interest. The animals are usually tested
during the day, at intervals of one to seven
days, and always after the apparatus has
been thoroughly cleaned with alcohol.
More details are given in Thiessen (1968).
Any significant deviation from this procedure will be indicated in the following
sections.

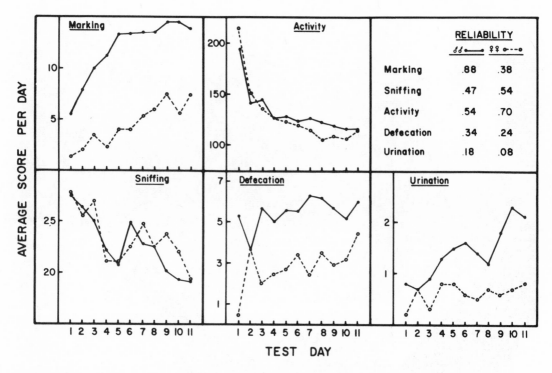

Figure 3.
Changes in ventral marking and other behaviors over time.

Preliminary observations indicated that there is a sexual dimorphism in ventral scent marking. Results from an early study involving ten males and ten females are shown in Figure 3 (Thiessen, Blum, and Lindzey 1969). Over an eleven-day period of testing, sexual dimorphism became exaggerated, with the males ventral marking, defecating, and urinating more frequently than the females. General activity and the frequency of sniffing the pegs did not differ significantly between sexes and adapted over the testing days. Interestingly, ventral marking correlated significantly with both defecation (r = 0.47) and urination (r = 0.50), suggesting that the three measures are a composite function serving the same territorial purposes. Sniffing frequency, on the other hand, correlated negatively with defecation (r = −0.50) and

urination (r = −0.45), whereas general activity was not significantly related to any other variable.

In this same study it was found that the pattern of scent marking changed over the eleven-day period. During the first few days of testing, both sexes marked more often during the final minutes of the testing period, but as they adapted to the apparatus the highest marking frequency shifted to the earlier minutes (Fig. 4). Apparently the sexual dimorphism is a stable characteristic of the gerbil, but marking frequency is partially contingent upon familiarity with the environment. Individual differences in responsiveness are very apparent in both sexes.

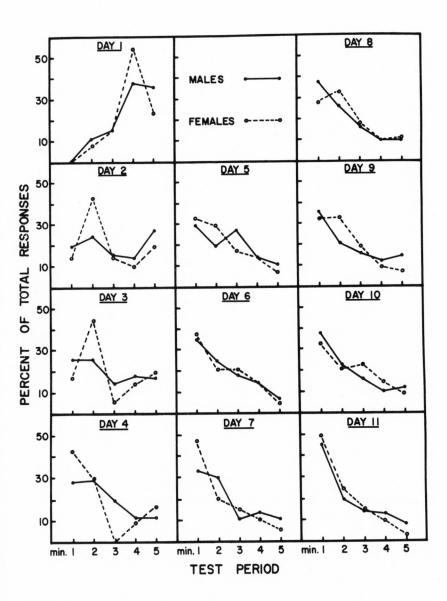

Figure 4.
Change in ventral marking pattern over eleven days.

Other general observations of males over a two-year period indicate that marking frequency is correlated with the season of the year, the age of the animal, and the number of tests given. These relations for the male are given in Table 10. In the laboratory ventral marking is most pronounced during September and October and is lower before and after the fall season. Independent of everything else, the oldest animals show the highest frequency of marking, and the more times the animal is tested, the more likely he will demonstrate high scores. All these variations are highly significant statistically.

Our experience is that gerbils are primarily nocturnal in activity. We supposed, therefore, that social activities, such as marking, would be more pronounced during the dark period. Two groups of ten gerbils were tested for marking during the day and night phase of a twelve-hour-on, twelve-hour-off light cycle in a day-versus-night counterbalanced order. Night observations were taken under dim red light.

The results for both groups clearly indicated that marking was higher at night than during the day. The first group scores were $\bar{X} = 2.4 \pm 1.40$ and $\bar{X} = 7.7 \pm 1.41$ for the day-night marking, respectively ($t = 3.19$; $p < .02$). The comparable scores for the second group were $\bar{X} = 1.5 \pm 0.51$ and 10.1 ± 2.73 for the light and dark phases ($t = 3.17$; $p < .02$). In other words, marking was from three to seven times more frequent at night. The order of testing, that is, day followed by night or night followed by day, had no obvious influence on the results.

All of these observations suggest that scent marking is a complex function involving sex, season, and diurnal variations. Marking, therefore, may be an indicant of the most vital social processes of the organisms. As such, the study of marking may reveal its relationship to territorial behavior and permit a fine-grained analysis of its physiological mechanisms. The extent to which we have been successful in our pursuit of this problem is discussed below.

Table 10.
Seasonal, age, and trial variations in ventral scent marking of the male gerbil

Month of Year

	Jan–Feb	Mar–Apr	May–Jun	July–Aug	Sept–Oct	Nov–Dec
Animals tested (N)	165	32	364	235	175	102
Marks/5-min trial ($\overline{X}$)	4.8	3.0	2.8	5.4	16.4	5.8

Age in Days

	100–125	126–150	151–175
Animals tested (N)	238	118	566
Marks/5-min trial ($\overline{X}$)	5.8	4.1	7.4

Number of Marking Trials (*Successive Weeks*)

	1	2	3
Animals tested (N)	555	555	555
Marks/5-min trial ($\overline{X}$)	4.2	5.1	5.7

Dominance, Population Density, and Scent Marking

For most mammalian species, gonado-tropic secretion is decreased and adreno-cortical output is increased at critical levels of population density (Thiessen 1964). Apparently social strife acts to switch the hormonal balance from that which is optimal for reproductive activities to that which is necessary to combat stress. We have predicted the same relationship between density and physiological processes of the gerbil.

To test this notion, sixty previously isolated males were assigned to differential living conditions in cages that measured sixteen by thirty by thirty-eight centimeters (Thiessen, Lindzey, Blum, and Wallace 1970). Twenty remained isolated, twenty were placed in four groups of five each, and twenty were placed in two groups of ten each. During the next eight days the groups were observed for fighting activity. Following this they were sacrificed and the following measures were recorded: body weight, number of wounds, ventral seba-ceous gland size in millimeters (length and width), and spleen, adrenal, seminal-vesicle, and testis weight. Body weight was determined to the nearest 0.1 gram and organ weight to the nearest 0.1 milligram.

While some fighting was observed in group cages immediately following the formation of the groups, it was neither pro-longed nor severe enough to cause wound-ing. The significant variations due to group-ing are seen in Table 11. Relative to iso-lated animals, grouped animals showed decreases in marking, deposition of urine pools, and seminal vesicle weight, and an increase in general activity. Other variables were not affected by grouping.

It would appear, therefore, that grouping does affect urination and ventral marking. Since seminal vesicle weight is generally considered to be an index of androgen secretion, the association between marking frequency and seminal vesicle weight is particularly interesting, suggesting an inter-action between marking and androgen titer. As the next section will indicate, androgen does indeed regulate the behavior. The fact that urination decreases with grouping again suggests that this is another method the gerbil uses to mark its territory. Physiological stress, as indexed by adrenal, spleen, and body weight, was not evident in this study.

Bell and Maier (personal communication) and Hull, Langan, and Rosselli (1973) have recently confirmed the depression of ventral marking as the result of grouping. In addition, Bell and Maier point out that one animal in the group continues to mark, presumably the dominant individual. Our data substantiate this finding (Thiessen, Lindzey, Blum, and Wallace 1970).

Table 11.
Effects of population density on marking and related measures

Measure	Social Conditions	$\bar{X} \pm SE$
Marking frequency (5 min)	Isolated (1)	16·95 ± 3·53
	Group (5)	9·90 ± 1·63
	Group (10)	6·55 ± 1·74
General activity (5 min)	Isolated (1)	139·00 ± 5·69
	Group (5)	158·45 ± 5·64
	Group (10)	162·70 ± 6·49
Number of urine pools (5 min)	Isolated (1)	1·00 ± 0·16
	Group (5)	0·40 ± 0·11
	Group (10)	0·50 ± 0·11
Seminal vesicle weight (g/100 g body weight)	Isolated (1)	529·60 ± 12·22
	Group (5)	496·15 ± 16·85
	Group (10)	464·95 ± 15·35

F test: $p < .05$.

Individual male gerbils were tested for marking frequency and were paired and observed daily for ten days for fighting and dominance during thirty-minute periods. Dominance was determined by a composite score for the ten bouts. On the last day of testing each gerbil was again assessed for marking behavior. During the next two days the animals were sacrificed and the following measures were obtained: body weight, ventral scent gland size, and testis, adrenal, seminal vesicle, and spleen weights. Body weight was determined to the nearest 0.1 gram and organ weights to the nearest 0.1 milligram.

The results can be easily summarized. Initial and terminal marking scores correlated significantly ($r = 0.77$), indicating behavioral stability regardless of the intervening fights. The difference in dominance scores between pairs correlated significantly with the difference in initial marking scores between pairs ($r = 0.38$), but not with the difference in terminal marking scores between pairs. Last, seminal vesicle weight correlated significantly with both initial marking and dominance scores ($r = 0.43$ and $r = 0.38$). All other relationships were insignificant. Thus, it appears that animals that are frequent markers and have high androgen titers, as evidenced by seminal vesicle weight, are also the ones most likely to win bouts. There was no evidence of physiological stress due to winning or losing (e.g., changes in body and spleen weight, adrenal weight, and urination), and no significant wounding was observed.

A similar effect was found when males met in triplet encounters during a two-week period. In this case grouping was continuous. Initially, severe fighting was observed in only eight of twenty group cages. One dominant animal invariably emerged in each of these eight cages, and no distinction could be made between the remaining submissive animals. Aggression normally

continued for several days, and in only two cases did the dominance order appear to reverse itself. In the eight cases in which fighting occurred, six deaths were recorded, two in one cage and one each in four others. Fighting and dominance variations were therefore available for seven groups; however, scores from isolated versus grouped conditions were available for all twenty groups.

The comparative results for dominant versus submissive animals and for isolated versus grouped animals are given in Table 12. Only significant values are indicated. The major associations with dominance were a high marking frequency, a decrease in ventral gland length, and an increase in ventral gland width. Total gland area, body weight, organ measurements related to stress, and other behavioral indices did not relate to dominance. The most pronounced and the only significant effects of grouping were decreases in general activity and relative testis weight and an increase in body weight. Absolute measures of spleen and adrenal weights were significantly elevated in grouped animals, but these organs did not vary between conditions when corrected for body weight differences.

Overall it is clear that dominance order and territorial marking are associated. Animals which mark frequently, and which apparently have high levels of blood androgen, have an advantage in aggressive encounters. Grouping did not always depress testis or body weights; however in all studies density decreased either seminal vesicle or testis weight, suggesting that grouping lowers androgen levels and possibilities for territorial dominance. Variations in findings among the studies may be due to specific housing conditions. Stress, as ordinarily defined, is not evident in group encounters, although it still may play a part in the overall effect. Studies to be cited later indicate that submissive animals can and will mark territories if the

Table 12.
Effects of dominance and grouping on marking and related measures

Measure		Social Condition	$\bar{X} \pm SE$
Dominant vs. submissive (N = 7 groups)	Marking frequency (10 min)	Dominant Submissive	39.00 ± 8.34 17.50 ± 5.82
	Change in gland length (mm)	Dominant Submissive	−5.00 ± 1.51 −1.93 ± 0.70
	Change in gland width (mm)	Dominant Submissive	0.71 ± 0.18 0.04 ± 0.20
Isolated vs. grouped (N = 20 groups)	General activity (10 min)	Isolated Grouped	317.00 ± 11.84 270.30 ± 8.34
	Terminal body weight (g)	Isolated Grouped	73.00 ± 2.93 77.60 ± 1.45
	Testis weight (g/100 g body weight)	Isolated Grouped	1,727.70 ± 43.65 1,593.30 ± 17.24

* *Test: p < .05*

intimidating influence of the dominant animal is removed. Whatever stress may be involved is evidently mediated in subtle ways or may require other techniques for measurement.

A male's continuing prerogative to mark in a social setting depends on his success in battle (Thiessen, Owen, and Lindzey 1971). This was demonstrated by placing two strange males, plus a female, in each of several living compartments and observing the marking, fighting, and chasing that occurred. The males were assigned to the pairs on the basis of their previous marking scores in the open field so that pre- and postinteraction scores could be compared. On the basis of original scores, high-markers were paired with high-markers, low-markers were paired with low-markers, and high-markers were paired with low-markers. In most cases, the males fought immediately and viciously, so that dominance was rapidly and decisively determined. High-markers usually defeated low-markers in mixed groups, but there was a stronger relationship between postinteraction marking scores and dominance than between preinteraction marking scores and dominance. Dominant males marked at high levels and submissive males marked very little, regardless of their previous marking scores (see part A of Table 13).

The same change in marking was observed when the dominant males competed across territorial boundaries. Following the initial competition, the dominant males remained in the compartments for two weeks; the partitions between pairs of adjacent living areas were then removed. The males quickly began to fight, and again the loser scarcely marked at all, despite his previously high marking scores. The winner, on the other hand, marked often, concentrating his marking in the loser's territory (see parts B and C of Table 13). Only after the newly acquired area was marked did the victor return to mark in his home compartment.

Clearly these data strengthen the contention that gerbil scent marking is related to territorial maintenance and acquisition. In diverse situations we have observed that male gerbils establish territories and defend them against intrusion by conspecifics. In addition, we have shown that a male's marking behavior changes as a result of this competition. Only socially dominant males scent mark appreciably, and they are quick to personalize newly acquired property. In general, only groups of males that are reared together are compatible as adults and defend their living area as a group. Our data suggest that olfactory cues from scent marking could be important signals of territorial possession. Evidence for this is presented in later sections.

Table 13.
Scent-gland marking during territorial disputes ($\bar{X} \pm$ SE for 30 min)

Group	N (Pairs)	Status Position Dominant	Submissive
A: Within-territory disputes variation			
1. High-low	18	23.9 ± 3.19	5.4 ± 1.09***
2. High-high	5	47.0 ± 10.25	14.0 ± 4.13*
3. Low-low	10	21.5 ± 5.79	5.4 ± 2.18*
B: Between-territory disputes variation			
Between territories	21	26.9 ± 7.50	2.0 ± 1.41***
Home territory	11	3.8 ± 1.75**	
Foreign territory	11	23.1 ± 4.83**	
C: Related to status change variation		Precompetition	Postcompetition
Within territory (low marker becoming dominant)	10	3.5 ± 1.73	23.7 ± 5.23**
Within territory (high marker becoming submissive)	12	22.2 ± 3.83	7.2 ± 2.35**
Between territories (previous dominant becoming submissive)	12	24.5 ± 5.79	1.2 ± 1.05**

t test: * $p < .05$
       ** $p < .01$
       *** $p < .001$

Olfactory Regulation of Behavior

The role of olfactory signals in gerbil social communication and in the regulation of scent marking has received much of our attention. Indeed, this has been one of the most challenging areas of inquiry. Gradually we are gathering information about the gerbil's responses to odor cues.

The effects of ventral gland excision, described in the next section, show that olfactory feedback of a male's own scent does not regulate marking behavior. Gland-less males displayed no deficit in marking relative to intact controls (Blum and Thiessen 1970). Furthermore, olfaction per se is not absolutely essential for marking to occur. Olfactory bulbectomy caused a

rapid and complete loss of marking behavior (Baran and Glickman 1970; Thiessen, Lindzey, and Nyby 1970); however, massive doses of testosterone (1,280 µg TP twice a week) partially reinstated the response (Thiessen, Lindzey, and Nyby 1970). These results are illustrated in Figure 5. The decrease in scent marking after olfactory bulbectomy apparently resulted from anosmia, not from neurological insult, because peripherally induced anosmia has the same effects (Wallen and Glickman 1974). Hence, both sensory input and sufficient hormone titers are necessary for the normal regulation of marking.

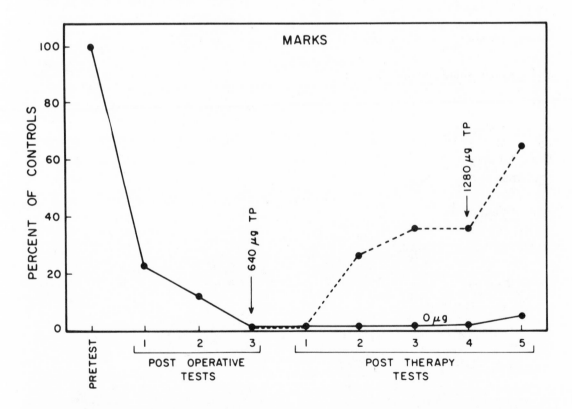

Figure 5.
Effect of olfactory bulb removal and TP supplements on marking frequency of male gerbils. The animals were tested forty-eight hours after the operation and again every fifth day.

Table 14.
Olfactory responsiveness of males and females to glandular sebum

Subject	Stimulus	Mean Time Spent Investigating Olfactory Stimuli (Min) ($\bar{X} \pm SE$)	t ($df = 29$)	p
Male	Male sebum	2.73 ± 0.39	3.44	<0.01
	Blank smear	1.25 ± 0.14		
	Female sebum	2.63 ± 0.28	1.26	NS
	Blank smear	2.29 ± 0.22		
	Male sebum	2.30 ± 0.26	0.48	NS
	Female sebum	2.53 ± 0.32		
Female	Male sebum	2.00 ± 0.18	0.44	NS
	Blank smear	1.89 ± 0.19		
	Female sebum	2.12 ± 0.31	1.28	NS
	Blank smear	1.67 ± 0.19		
	Male sebum	2.02 ± 0.20	1.69	NS
	Female sebum	2.45 ± 0.31		

A gerbil's reaction to conspecific odors depends on sexual and social parameters. Using two different techniques, we have shown that males are initially attracted to scent gland odors of other males. This was demonstrated first using a Y-tube choice apparatus (Thiessen, Lindzey, Blum, and Wallace 1970). Paper strips smeared with sebum or no sebum (skin smear) were taped behind mesh screens in each of the two goal boxes. Then male and female gerbils were individually tested for the time spent in each goal box. Both sexes were given choices of (a) male versus female sebum, (b) male sebum versus no sebum, and (c) female sebum versus no sebum. As shown in Table 14, males were attracted to male odors and females showed no preference. The males' preference disappeared when the choice was male versus

female sebum, although female gland odors did not seem to attract the males. The estrous condition of the donor females was unknown and may account for this discrepancy. The portion of this study using male sebum versus no sebum was replicated using all male subjects. The males consistently preferred the gland odors of other males.

We obtained similar results with a technique which we call a sniff test. For this procedure a rectangular block of wood was hung in a gerbil's home cage. Embedded in the wood were two cups, each filled with paper and covered by a thin sheet of tissue paper and a wire mesh screen. The paper in one cup was impregnated with sebum from a male or female; the paper in the other cup was clean. An experimenter observed the gerbil's attraction to the cups

Table 15.
Habituation of males' attraction to male sebum

	Test 1	Test 2	Test 3	Test 4
Correct identifications of sebum cup based on male's interest	10/11 (91%) $p < .008^*$	5/11 (46%) $p < .726$	4/11 (36%) $p < .885$	4/11 (36%) $p < .885$

Note: Males tested received sham olfactory bulbectomies between Test 1 and 2.
   * One-tailed binomial probability test, with correction for continuity.

for sixty seconds and then predicted which cup contained the sebum. Another experimenter rotated the cups in a random fashion. The observer never knew the true position of the sebum cup.

The sniff test was originally used as a test of olfactory sensitivity in the olfactory bulbectomy study discussed above. The original thirty-three male subjects and their mates were individually tested for their attraction to male sebum versus no sebum prior to any experimental manipulation. As in the Y-tube, males were attracted to male sebum (26/33 or 79 percent correct predictions, $p < .001$) and females were not (20/33 or 61 percent, $p < .147$). Following olfactory bulbectomy or sham operations, the males were retested, but the sniff test was inadequate for distinguishing the groups. Even the sham-operated males no longer showed a preference, as detailed in Table 15. Apparently their interest in the odor had habituated since the previous test.

The sniff test itself, at least for the first trial, seems quite reliable. To determine if our prediction criterion was appropriate, we tested 154 males for their attraction to either male or female sebum versus no sebum. Each prediction was rated as: *positive*—the gerbil definitely preferred one cup to the other and sniffed at it intently; *clue*—the gerbil displayed more interest in one cup than the other, but the preference was weak; or *guess*—the gerbil displayed no interest in either cup or displayed equal interest in both. Table 16 shows that the percentage of correct predictions corresponded closely to the prediction ratings. It also shows that males exhibited stronger attraction to female sebum than in the Y-tube study, but male odors still elicited a larger response. Again, the estrous cycle of the females was unknown. The predictions themselves were equally accurate for both types of stimuli.

Table 16.
Correct identifications of sebum cups based on male's interest

| Stimulus | Prediction Rating* | | | Total |
	Positive	Clue	Guess	
Male sebum	6/6 (100%) p < .021**	35/51 (69%) p < .006	9/21 (43%) p < .803	50/78 (64%) p < .009
Female sebum	5/5 (100%) p < .036	29/45 (64%) p < .036	14/26 (54%) p < .42l	48/76 (63%) p < .015
Total	11/11 (100%) p < .0015	64/96 (67%) p < .001	23/47 (49%) p < .618	98/154 (64%) p < .0006

* *See text*
** *One-tailed binomial probability test, with correction for continuity.*

The habituation of the initial orientation response was re-examined using the sniff test. This time males were tested for their attraction to male or estrous female sebum versus no sebum. The female donors were brought into estrus by injections of estradiol benzoate (EB) and progesterone (P) (100 μg and 200 μg, respectively) before the ventral gland smears were taken. The next day, half of the males exposed to male sebum interacted with their respective sebum donors for fifteen minutes in the open field. Likewise, half of the males exposed to female sebum interacted with their respective female donors. The remaining males were placed alone in the open field for fifteen minutes. The sniff test was repeated the following day. One week later, and again two weeks later, the males interacted with the same donors and were retested for attraction to their odors the next day. The female donors were reinjected with EB and P prior to the second and third interaction periods. Table 17 indicates that males were attracted to sebum from estrous females just as they were to male sebum. The initial orientation to either odor habituated over trials, though the intervening periods of social contact with the sebum donor helped to maintain the male's interest. This was most evident with estrous female donors.

The attractive odors of male sebum may be testosterone-dependent, as is true for the gland itself. However, the quality of the odor does not appear to change within forty-eight hours after castration. We castrated several males two days before using them as donors for the sniff test. Males exposed to sebum from these castrates versus no sebum preferred the castrates' odors (17/24 or 71 percent, p < .033). This is very close to the usual level of preference for normal sebum.

Presumably, the hormonal state of a gerbil could also affect its responses to the scent gland odors of conspecifics. Preliminary data, though, suggest that this is not the case. Male gerbils were tested before and again several weeks after castration for attraction to male sebum versus none. On these respective tests their preferences were 8/11 or 73 percent (p < .116) and 11/11 or 100 percent (p < .002). Their postcastration responses were comparable to those of intact males, although their own initial preference was somewhat low. Baran (1973) has also shown that both intact and castrate male gerbils are attracted to odors (soiled bedding) from strange males and that at least some of these attractive odors are produced by castrates.

A small study with females yielded similar results. Isolated, ovariectomized females were injected with EB and P to induce estrus or they received oil injections. Then each group was tested for attraction to male sebum versus no sebum. Several weeks later, the hormonal treatments were reversed for the two groups and the females were retested. The data are given in Table 18. There is no indication that the estrous state affected their responses. It is curious, though, that, on the first test, both groups of isolates displayed a greater attraction to male odors than did mated females in an earlier study (see above). In the Y-tube study, isolated females were not attracted to male odors at all. However, the females in the Y-tube study were sexually naïve, while the isolates in this sniff test were not. Perhaps the female's attraction to male odors varies with sexual experience and the presence of a mate.

Table 17.
Attraction of males to male or female sebum following social interaction

Stimulus	Social Experience	Pretest	Test 1	Test 2	Test 3
Male sebum	Interaction with male	17/25 (68%)* p < .055**	16/25 (64%) p < .116	16/25 (64%) p < .116	15/25 (60%) p < .212
Male sebum	No interaction	17/25 (68%) p < .055	19/25 (76%) p < .009	10/24 (42%) p < .850	12/24 (50%) p < .579
Female sebum	Interaction with female	18/25 (72%) p < .023	14/25 (56%) p < .345	22/25 (88%) p < .0002	17/24 (71%) p < .033
Female sebum	No interaction	17/25 (68%) p < .055	17/25 (68%) p < .055	17/24 (71%) p < .033	13/24 (54%) p < .421

* Correct predictions of sebum cup based on male's interest.
** One-tailed binomial probability test, with correction for continuity.

Table 18.
Attraction of estrous and nonestrous females to male sebum

Subjects	Test 1	Test 2
Estrous female	10/11 (91%) p < .008*	7/10 (70%) p < .187
Nonestrous female	9/11 (86%) p < .036	7/10 (70%) p < .187

* One-tailed binomial probability test, with correction for continuity.

Conspecific odors also alter male marking behavior, particularly following aggressive interaction. An early report of gerbil scent marking indicated that male gerbils seemed to mark less in an area marked by other males than in a clean one (Thiessen, Blum, and Lindzey 1969). This fit the concept of territorial signaling, but, when a systematic study was attempted, males marked clean and contaminated areas to the same degree (Thiessen, Lindzey, and Nyby 1970). Any hesitance observed in the first study may have been due to conflicting investigatory responses, rather than to avoidance of a previously marked area.

The situation is quite different once the odor cues have acquired social significance. We have evidence that males who have been defeated by a group of gerbils show a decrease in marking in the presence of the group's odors (Nyby, Thiessen, and Wallace 1970). In this study two groups of males were tested daily for marking in half of a large living area, while the resident gerbils were confined to the other half. For one group, the residents could enter the test compartment after each test and attack the intruders. These attacks were immediate and severe. The control group never interacted with the residents. The differential changes in marking that resulted from the aggressive interactions are depicted in Figure 6. Clearly, defeat suppressed the male's tendency to mark. After ten encounters marking was depressed 75 percent relative to control animals. Furthermore, Figure 6 shows that the suppression of marking was associated with olfactory cues from the resident and defeated animals. The defeated males marked at control levels in the open field or in the test compartment if it was thoroughly cleaned. Yet their marking in the open field declined when contaminants (e.g., dirty sawdust) from the residents' area were present. Testis, adrenal, seminal vesicle, and spleen weights were normal in the defeated males,

ruling out long-term hormonal changes as the mediating variable. In other words, scent signals may initially act as attractants, but in order for them to continue to have significance they must be conditioned to social features in the environment.

Defeat also changed the males' responses to the residents' odors in the Y-tube preference test. Given a choice between wood shavings from their home cage or from the residents' territory, defeated males spent more time near their own odors. Control males showed no preference. Either defeated males avoided the residents' odors or they preferred their own.

Thus we can conclude that male gerbils are not initially intimidated by the odors of other males. On the contrary, they are attracted to these smells. But, following aggressive interactions, dominant males' odors can suppress scent marking by subordinates and may, in fact, repel them. Further study will be required, though, before the social significance of gerbil scent marks can be clarified. In an attempt to replicate and extend the observations on social subordination and scent marking, Yahr (1972) found that defeated males would not scent mark in the defended territory even after it had been cleaned. They would, however, scent mark normally in a neutral arena even when odors associated with defeat were present. Visual cues seemed to be more prominent in this study than in the previous one. Perhaps gerbils use visual cues when they are available yet rely on olfactory cues at night and underground.

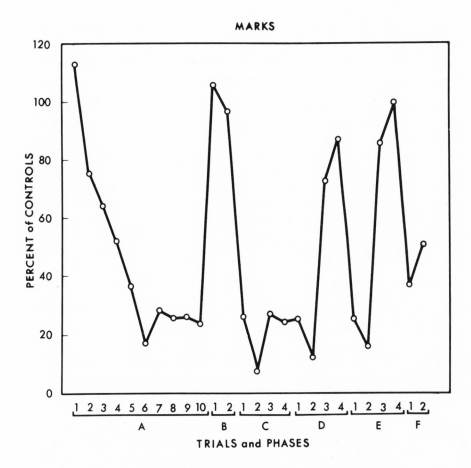

Figure 6.
Effects of social interaction and sensory cues on marking frequency of male gerbils. *Phase A:* Decrease in marking following social defeat. *Phase B:* Recovery of marking in a clean, unfamiliar apparatus. *Phase C:* Suppression of marking by olfactory cues from other defeated gerbils. *Phase D:* Recovery of marking in residents' territory when all olfactory cues are removed. *Phase E:* Recovery of marking in the resident's territory when auditory, but not olfactory cues were present. *Phase F:* Suppression of marking in a novel apparatus by olfactory cues from residents.

The social and olfactory parameters that regulate female marking are unknown, but a recent study suggests that females use sebum odor to recognize pups (Wallace, Owen, and Thiessen 1973). On various occasions we have observed females marking their pups; this may be an important aspect of their maternal behavior. In one study each of eleven individually housed, lactating females was presented with two unfamiliar pups outside her nest area. Both pups had been cleaned with ethanol and one was smeared with sebum from the test mother's gland. Ten lactating mothers retrieved at least one pup into the nest, and seven retrieved the sebum-smeared pup first. When the test was repeated, using a slightly different technique, eight mothers retrieved, and all retrieved the marked pup first. Pups not smeared with sebum were eventually retrieved, but it took longer to bring them to the nest. Incidentally, of the eleven times the females were observed to scent mark a pup, all marks were directed at the pup not smeared with sebum.

Twelve-day-old gerbil pups are, in turn, attracted to the odors of their home cage. They move toward and settle down near cage shavings from the mother's cage in preference to clean sawdust or cage shavings from strange lactating or nonlactating females (Wallace, unpublished). Maternal scent marking may be very important for the mutual identification and attraction of the mother and her young.

We are now developing techniques to study the animals' ability to respond to sebum in order to obtain food. In one task, on which we have considerable data, males must discriminate between sebum smears on filter paper and clean filter paper in order to successfully dig for a food reward. Although training is still in progress, eight animals can make a perfect discrimination on every trial.

Thin-layer chromatographic separation of the whole sebum, using a benzene-acetone (80–20) solvent, reveals up to twelve separate bands when the plate is sprayed with sulfuric acid and heated. The odor can be extracted from the band area with the approximate Rf (position of band relative to solvent front) of 0.35 and is readily identifiable as the musk odor deposited by the marking behavior. When animals are trained to suppress bar pressing for food when whole sebum is introduced into an air stream, they also suppress bar pressing to the material extracted from the TLC (thin layer chromatography) plate at Rf 0.35 but not to material scraped from other areas of the plate. Spontaneous exploration of this fraction in an open field is also high.

Fatty acid extractions using sodium hydroxide (NaOH) or sodium bicarbonate (NaHCO$_3$), coupled with gas chromatography and mass spectrophotometry have allowed us to identify an active biological component as phenylacetic acid (Thiessen, Regnier, Rice, Isaacks, Goodwin, and Lawson 1974). This acid runs conjointly with the active fraction on the TLC plate and suppresses bar pressing as well as does the sebum fraction. Similarly, phenylacetic acid stimulates exploration. Other preference tests now underway suggest that phenylacetic acid acts like whole sebum in directing an animal's attention and stimulating marking.

Olfactory cues and scent marking obviously serve a variety of functions for the gerbil. Attraction and aggregation may be the primary roles of sebum odors, particularly for males. Later, when territorial competition is complete, other responses may predominate. Halpin (1974) has found that gerbils can discriminate between individuals on the basis of sebum cues. Apparently sexual and social status, as well as individual recognition and food recognition, can be communicated by the scent gland cues.

Reproductive Isolation

The evolution of genetically distinct populations is commonly thought to require reproductive isolation of sufficient duration for the accumulation of genetic differences. The antecedents of the concept of reproductive isolation extend back to Aristotle. Isolation became a primary concept in Charles Darwin's demonstration of natural selection and was elaborated on by Dobzhansky (1937) and Mayr (1970). The topic is still of great interest, as the exact mechanisms leading to reproductive isolation are not clear.

Recent evidence suggests that social behavior can be important in initiating isolation between gene pools. John J. Christian (1970), for example, has suggested that as population density increases some animals are forced to migrate from the preferred habitat into unoccupied areas. This could lead to isolation from the core group and genetic differentiation. Miklos Udvardy (1970) contends, however, that migratory animals must become geographically isolated from the main group in order to prevent reciprocal gene flow. In other words, Christian's model could account for reproductive isolation only if it could be demonstrated that socially induced migration is across a geographical barrier.

We have devised a model situation in our laboratory which may help explain how submissive animals become isolated (Thiessen and Dawber 1972). Essentially what we have done is to introduce male and female intruders one at a time into a territory that is occupied by a resident pair and provide the intruders an opportunity to swim across a water barrier (an aquarium) to a safe unoccupied territory. A schematic version of this experiment is presented in Figure 7.

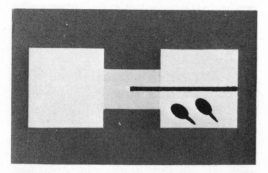

Phase 1

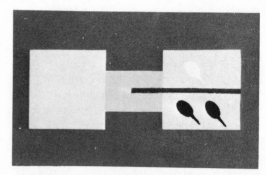

Phase 2

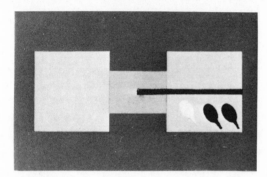

Phase 3

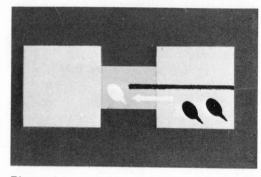

Phase 4

Figure 7.
Schematic presentation of the effects of social pressure on a gerbil's tendency to migrate across an ecological barrier. *Phase 1:* A pair of gerbils establish residency in one compartment. *Phase 2:* An intruder gerbil is adapted to another compartment and observed for migration. *Phase 3:* The intruder is introduced into the residents' territory. *Phase 4:* The intruder is observed for migration following attack by the residents.

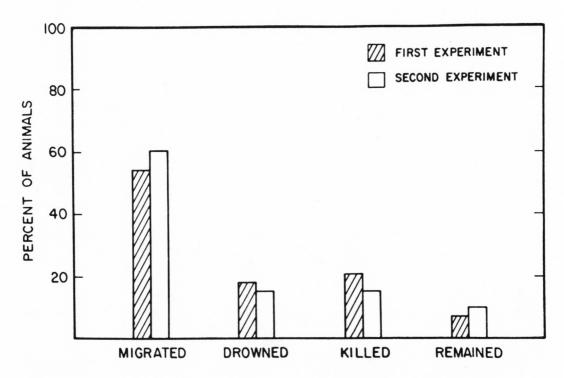

Figure 8.
Percentages of intruders migrating, drowned, killed, and remaining in an occupied territory.

We have found that the test animals did not swim unless provoked by territorial aggression. When the intruder was exposed to the resident pair, fighting broke out immediately. In two separate experiments involving forty-eight intruders, approximately 58 percent migrated within twenty-four hours, 16 percent drowned in their attempt, 18 percent were killed before they could exit, and only 8 percent remained in the residents' territory. Those that remained were still under attack after twenty-four hours. The data are summarized in Figure 8. Clearly, territorial aggression is a potent impetus for migration. Our evidence suggests that the migrants would never mate in the occupied territory and would remain isolated.

Olfactory cues from the original territory may reinforce isolation even if the intruders are inclined to swim back. In the second migration experiment, we found that social migrants avoided odor from the residents' territory. In this case migrants were given a choice between a cage containing soiled shavings and feces from the residents' territory and a cage that contained clean shavings. The interconnected cages were constructed in such a way that the animals could not come into physical contact with the shavings, but could only smell them. Under these conditions, nine out of eleven migrants who built nests did so in the cage containing clean shavings, and three that did not build nests showed no clear choice. Olfactory cues may therefore reinforce avoidance behavior and help insure that reproductive isolation is complete. This is another example of how scent signals take on significance after critical social interactions.

4

Hormone Regulation of Territoriality in the Gerbil

The acquisition and maintenance of territorial rights are often regulated by testicular hormones (Carpenter 1958; Crook 1968; Ewer 1968; Lack 1966; Tinbergen 1957; Wynne-Edwards 1962). Usually the onset of territoriality is associated with puberty and with the recrudescence of the testes at breeding.

While it is generally accepted that hormones, especially testosterone in the male, act in territorial processes, it is not clear how they act. Few systematic studies have been conducted, and most of our information takes the form of correlations between territorial acquisition and morphological changes in the gonads. For territorial scent marking the deficit of information is even greater. With the exception of the work on the rabbit (*Oryctolagus cuniculus*) by Mykytowycz (1965) and the Columbian ground squirrel (*Spermophilus columbianus*) by Kivett (1975), there is little evidence indicating a relationship between territorial behavior, scent marking, and hormone mechanisms.

Our own studies summarized in this chapter are meant to clarify the mechanisms of action by which the gerbil's territorial behavior is expressed through ventral scent marking. Inferences about hormone control have already been made in the earlier section on social behavior. In this section several specific mechanisms are implicated. Castration and hormone-replacement experiments are reported for the male and female, some developmental data are given for the male, and blood testosterone measures are reported for the male. Overall it is becoming apparent that scent marking and territorial behavior are under strict control of specific hormones. In the following chapter the way in which these hormones affect the central nervous system is outlined.

Basic Hormone Patterns in the Male

The first critical observation on the androgen control of scent marking in the gerbil was in 1967, when we found that castration of the male led to a rapid decline of marking and a total disappearance of the ventral sebaceous gland pad (Thiessen, Friend, and Lindzey 1968). In this experiment, marking frequency was first recorded in the usual way in an open field studded with Plexiglas marking pegs. This was followed by castration and then marking tests every three days.

As seen in Figure 9, relative to sham-operated controls, marking fell off rapidly following castration. Ventral gland size (length times width in sq cm) also declined, and eventually the gland disappeared. After approximately three weeks both measures were at a minimum.

MARKING

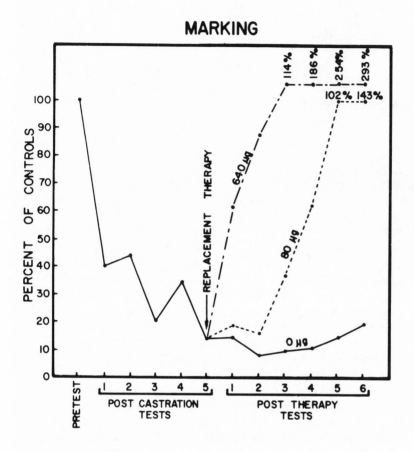

Figure 9.
Effects of castration and TP replacement therapy on territorial marking (*left*) and ventral gland size (*right*).

Thereafter, three groups of castrates were formed: one group received 640 micrograms of TP, a second received 80 micrograms of TP, and a third received only oil injections. All hormones were administered subcutaneously every three days. As is obvious in Figure 9, both scent marking and gland size responded to TP in a dose-dependent manner. Both of the measures recovered, and marking reached nearly 300 percent of that for controls with the higher dosage of TP. The size of the gland pad was more resistant to TP and barely exceeded control values.

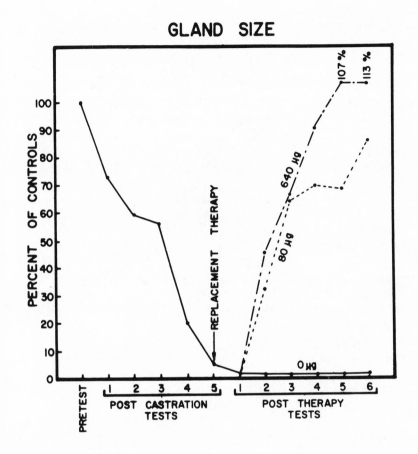

Blum and Thiessen (1971) have extended this experiment, obtaining essentially the same results. In this case, however, an ascending series of TP doses were administered to different groups of castrates once a week for seven weeks. As before, and as indicated in Figure 10, castration resulted in a substantial fall in scent marking. When different groups of these castrates were given 0, 32, 64, 128, or 256 micrograms of TP, the recovery of marking was clearly dose-dependent. Once more the largest dose of TP stimulated "super" levels of marking. Changes in gland size paralleled changes in marking, but again the gland appeared more resistant to the hormone and did not exceed normal levels.

In the same experiment correlations were obtained between seminal vesicle weight and marking frequency in a separate group of intact males. The correlation was 0.54, indicating that individual differences in androgen titer, as measured by seminal vesicle weight, accounted for a substantial portion of the individual differences seen in marking frequency. These data and those from the castration experiments strongly implicate the testes in the regulation of territorial marking and ventral gland secretion. Marking is much more sensitive to hormone change than is the morphology of the scent gland, but both respond in a similar fashion. Only marking, however, can be stimulated much above control levels.

The opposite is true, though, for males that do not scent mark spontaneously (Yahr in press *b*). These males do not scent mark when castrated and injected with TP, but their scent glands enlarge. Their glands are initially smaller than those of spontaneously marking males (1.12 ± 0.4 vs. 1.32 ± 0.6 sq cm), but androgen therapy eliminates this difference. Since the scent glands of zero-marking males are sensitive to testosterone, their normally small size suggests that endogenous androgen titers in these males are low. However, low hormone titers can not account for the absence of scent marking. Perhaps zero-marking males are behaviorally insensitive to androgens as a result of inadequate hormone stimulation earlier in their development.

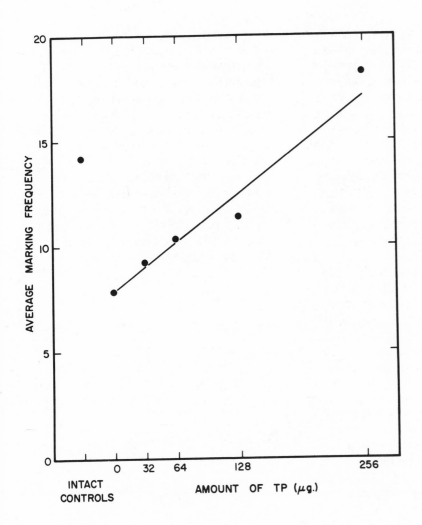

Figure 10.
Relationship between male marking scores and amount of TP injected weekly. The calculated regression of marking frequency on dose level is shown by the line.

As expected, the ontogeny of marking in the male is hormone-dependent (Lindzey, Thiessen, and Tucker 1968). When males were castrated prepubertally at thirty days of age, neither marking nor ventral gland growth occurred in later life. Ordinarily marking and substantial gland growth occur around seventy to one hundred days of age. When, however, 640 micrograms of TP were administered every six days beginning at about fifty days of age, intact animals began marking by sixty-five days of age and outperformed controls by 200 percent at one hundred days of age. These variations are seen in Figure 11.

Consistent with the earlier results, hormone supplements did not increase ventral gland size beyond that of intact controls. In fact, the gland failed to respond at all until marking had already surpassed control levels by substantial amounts. These and other experiments indicate that territorial marking and gland secretion are broadly associated, but not in a one-to-one fashion as would be expected if they were causally related.

There are additional indications that testosterone in the male is the hormone controlling scent marking (Rice 1975). First, dihydrotestosterone is not evident in the plasma of young or adult males. Second, when developmental changes in plasma testosterone were related to scent marking and other indices of sexual development, some obvious correlations appeared. All indices of androgen secretion increased between the ages of 40 and 100 days, with some measures leveling off or decreasing by 150 days. Testosterone titers generally increased after 40 days of age but showed a great deal of variation (0.5–2.5 ng/ml). The ontogeny of ventral scent marking approximated earlier observations, with a substantial increase around 90 days of age.

Interestingly, testosterone levels did not reflect marking frequency as closely as seminal vesicle weights, although the behavioral test most closely associated with blood sampling showed the strongest relationship (several tests for marking were conducted over a period of days). The impression gained from this study was that short-term fluctuations in testosterone did not reflect the average developmental status of the hormone-dependent behavior, as did more stable secondary sex characteristics. Blood testosterone levels, scent marking, and gland size are not necessarily related at any particular moment.

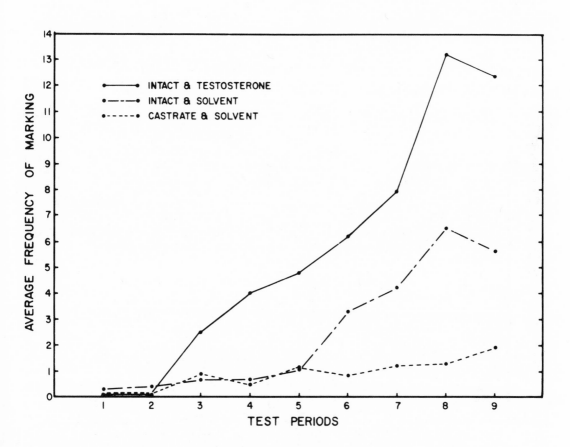

Figure 11.
Development and hormonal control of territorial marking in the male gerbil. Testing began at fifty-two and ended at one hundred days of age.

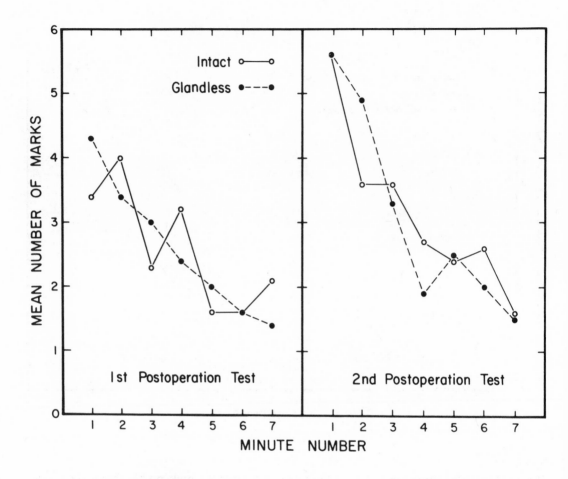

Figure 12.
Changes in marking frequency over a seven-minute period for glandless and intact gerbils.

This point is supported by an experiment in which the midventral gland of intact males was excised (Blum and Thiessen 1970). Even after the total loss of the marking gland, marking continued over sequential trials and in two separate tests at the same frequency and pattern as that observed in sham-operated controls. The data are seen in Figure 12. Hence, the functional regulation of marking is not controlled by peripheral cues from the gland and must be related to central processes. This conclusion, though, must be qualified by the information discussed earlier which indicates that olfactory cues from other animals do indeed modulate territorial functions, including marking frequency. While normal levels of marking are modified by olfactory cues, it is still clear that the gland need not be intact in order for marking to occur.

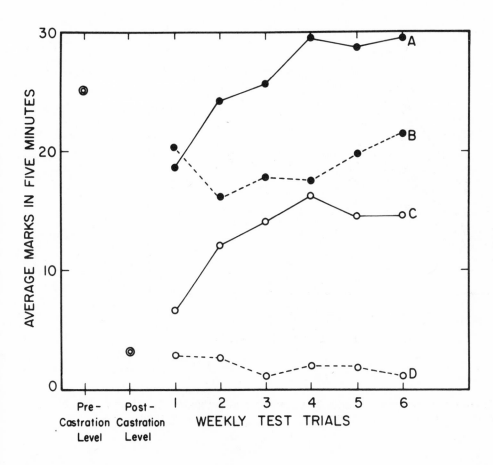

Figure 13.
Effects of castration and hormone treatment on marking frequency of males. Condition:
A = TP, 300 μg/week; B = TP + EB, 300 μg + 160 μg/week; C = EB, 160 μg/week; D =
oil control.

Subsequent work shows that the male is responsive to estrogen as well as testosterone (Nyby and Thiessen 1971). Males that were castrated showed the usual drop in marking and the usual increase following weekly injections of 300 micrograms TP. Estradiol benzoate given to castrates at the rate of 160 micrograms a week also elevated marking, although not to the same extent as did the same dose of TP (Thiessen, Friend, and Lindzey 1968). Other castrates received TP and EB combined. Their marking increased but remained

below that of animals receiving TP alone. The results are given in Figure 13. Ventral gland weights and gland smear weights (obtained by rubbing the gland with filter paper) did not differ among the hormone conditions, and all were heavier than for oil-injected castrates.

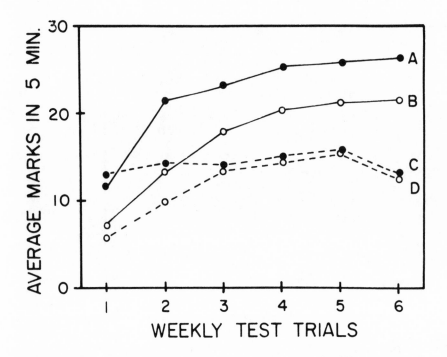

Figure 14.
Effects of hormone treatment on marking frequency of intact males. Conditions: *A* = oil control; *B* = P, 800 μg/week; *C* = P + EB, 800 + 80 μg/week; *D* = EB, 80 μg/week.

Apparently, marking in males is responsive to estrogen as well as testosterone, but not to the same degree. In combination, EB and TP may compete for the same receptor sites. These results are very similar to those involving sex behavior in the rat, where male behavior can be activated by estrogen but not to the same degree as by testosterone (Ball 1939; Davidson 1969). In contrast to the differential effects of EB and TP on marking, both hormones are capable of stimulating ventral gland secretion. Work to be presented in the next section verifies that marking is under more rigorous molecular control than is glandular secretion.

When intact males were administered the same amount of EB (80 μg twice a week) or a combination of EB and P (80 μg and 800 μg, respectively) marking was sharply depressed (Thiessen, Lindzey, and Nyby 1970) (see Fig. 14). Progesterone alone was ineffective, and the two hormones did not interact. The mode of action for this depression was apparent when the testes and seminal vesicles were weighed at the conclusion of the experiment: estrogen but not progesterone severely decreased these organ weights, indicating that androgen secretion had been curtailed. These data are seen in Figure 15.

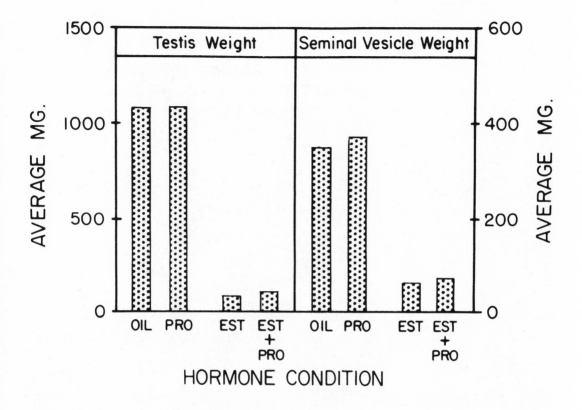

Figure 15.
Effects of hormone treatment on testis and seminal vesicle weights. Conditions: *oil* = oil control; *pro* = progesterone, 800 μg/week; *est* = estradiol benzoate + progesterone, 80 + 800 μg/week.

It is probably the case that moderate levels of marking were supported by the estrogenic steroid, as discussed earlier for castrates, and that little androgen was present. Histological examination of the testes revealed tubular atrophy, absence of spermatogenesis, and an apparent lack of Leydig cells (see Fig. 16). Estrogen may reduce the secretion of pituitary gonadotropins necessary for testicular activities.

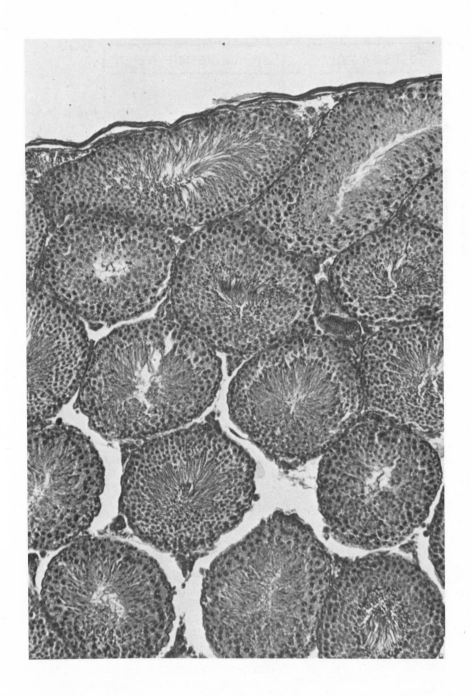

Figure 16.
Effects of estrogen on seminiferous tubules and spermatogenesis. *Left:* normal testis; *right:* testis from an estrogen-treated gerbil showing tubular atrophy and absence of spermatogenesis.

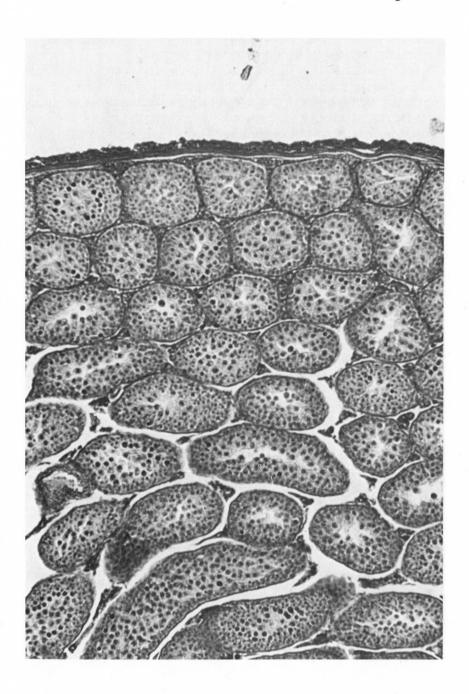

Males also chin objects in their environment (Thiessen, Yahr, and Lindzey 1971) with granulated sebaceous tissue located in the neck and chin area. The conditions under which chinning occurs are rather obscure at the moment, but one experiment showed that males chin high objects that cannot be skimmed with the ventral marking gland. Supplements of 640 micrograms TP weekly did not increase chinning over normal levels, although it is still possible that the normal range of chinning is under androgen control. Chinning of objects may have similar signaling qualities to ventral marking, and it offers an alternative way to apply pheromones to surrounding objects. Our general observations indicate that gerbils rub their necks down the sides of tall objects and around doorways. This may be another way in which they personalize their environments. Gerbils also sniff intently in the chin area, the ventral gland area and the anogenital region of strange gerbils upon meeting and when they are reunited with relatively familiar animals. Olfactory cues from each of these areas may be involved in individual recognition among conspecifics.

The status of a male gerbil's scent gland also depends on the amount of water he consumes. Although gerbils can survive without water, maintaining fertile pairs of gerbils on metabolic water (a weekly ration of lettuce was the only water source) stopped their reproduction and in the males reduced scent gland size, relative testis weight, and relative seminal vesicle weight (Yahr and Kessler 1975). Males in this situation also sired fewer young, even if the females had access to water. Bannikov (1954) reports that gerbils breed during spring and summer, when the rainfall is heaviest, and eat green plant pulp rather than seeds at this time. The seasonal changes in water consumption that result may affect testicular functions, thus triggering annual changes in the gerbil's mating and territorial behavior.

In studying hormone control of territoriality in gerbils, the role of androgens in regulating aggression has been difficult to determine. According to Sayler (1970), gerbils are similar to other mammals studied in that aggression declines after castration and increases again after TP therapy. Interestingly, though, other researchers report that castrated male gerbils fight more often than gonadally intact males or castrates treated with TP do (Anisko, Christenson, and Buehler 1973; Christenson, Wallen, Brown, and Glickman 1973). The basis for this discrepancy is not clear.

To add to the controversy, a more recent study found that castrates and intact males were equally likely to fight under several conditions, though individual differences were large (Yahr, Coquelin, Martin, and Scouten, unpublished). One consistent difference between the groups, however, was that intact males often killed their opponents, whereas the castrates rarely did. When the males fought in their own territory, both intact and castrate residents tended to dominate intruders, but this advantage was more prominent among the intact males.

Basic Hormone Patterns in the Female

Ventral marking in the male depends on testicular androgens. It is becoming equally clear that the gonads also regulate marking in the female. An initial investigation indicated that intact females respond to TP (640 μg twice a week) with increases in scent marking, but that EB (80 μg), P (800 μg), and growth hormone (2.5 IU) were ineffective (Thiessen and Lindzey 1970) (see Table 19). Neither gland size nor secretion responded to any hormone except TP, and, interestingly, defecation, which may be another index of territorial marking, was elevated by TP. On the basis of this experiment it is obvious that females can be induced to scent mark at unusually high levels by testosterone, but it is less clear if estrogen and progesterone have any effects.

Table 19.
Effects of hormone supplements on territorial responses of the female Mongolian gerbil

Hormone Condition	Test Scores (Mean ± SE)			
	Marking Frequency (per 5 Min)	Gland Size, Length times Width (in Sq. Cm)	Gland Secretion (Weight of Blot in Mg)	Defecation (Boluses per 5 Min)
Vehicle	5.00 ± 2.91	0.605 ± 0.073	1.94 ± 0.24	3.08 ± 1.16
Testosterone	25.17 ± 3.15	1.069 ± 0.049	3.55 ± 0.45	7.42 ± 1.28
Estrogen	3.67 ± 1.53	0.641 ± 0.070	2.42 ± 0.37	5.83 ± 1.55
Progesterone	0.42 ± 0.34	0.577 ± 0.033	1.89 ± 0.24	2.42 ± 1.14
Somatotrophin	2.25 ± 1.35	0.651 ± 0.060	2.41 ± 0.40	5.92 ± 1.20
F	22.63, $p<.001$	11.97, $p<.001$	3.63, $p<.05$	2.72, $p<.05$

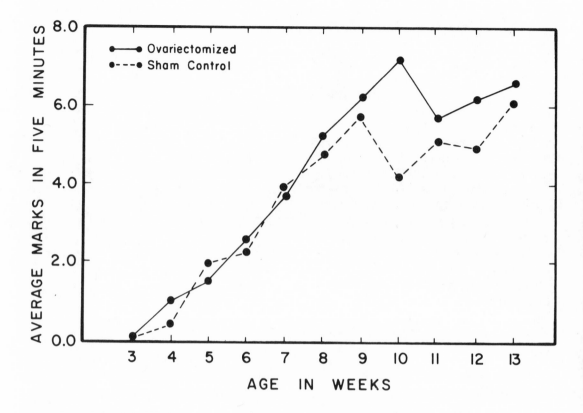

Figure 17.
Development of female marking behavior in ovariectomized females and sham-operated controls.

In a subsequent study it was found that ovariectomy at twenty-two days of age did not affect the onset or the developmental pattern of marking (see Fig. 17). And ovariectomy of adult females did not result in a depression of marking, even though the scent gland disappeared (Whitsett and Thiessen 1972). Similarly, it was found that neither ovariectomy, adrenalectomy, nor a combined operation disrupted marking in adult females (Thiessen, Owen, and Lindzey 1971).

It was tempting to conclude at this point that females respond to the heterotypical hormone, testosterone, with increases in marking, but that ovaries (and adrenals) do not contribute hormonally to the usual levels of female scent marking. It is important to note that the usually low levels of marking observed in females correspond closely to the levels of marking found in males following castration. Perhaps both sexes display a base level of marking that is not influenced by the gonads.

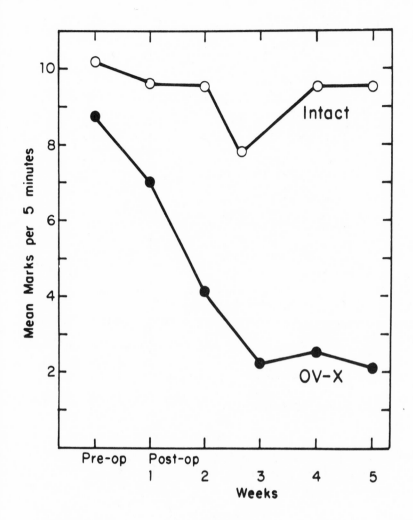

Figure 18.
Effects of ovariectomy on the marking frequency of high-marking females.

Nonetheless, recent information shows that ovariectomy does indeed decrease marking in females that show high levels of scent marking preoperatively (Wallace, Owen, and Thiessen 1973). As can be seen in Figure 18, adult females that marked approximately ten times per five minutes decreased their frequency of marking approximately 80 percent following ovariectomy. Similarly, postpartum ovariectomy substantially reduced the high levels of ventral marking associated with lactation (Fig. 19).

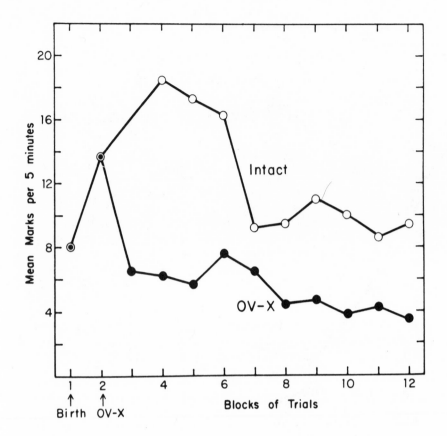

Figure 19.
Effects of lactation and ovariectomy on territorial marking in females.

Owen (1972) has extended these experiments by showing that large doses of estrogen (160 μg/week) or progesterone given singly to ovariectomized animals did not induce marking. However, when the two were given in sequence—EB followed by P thirty-six hours later—then certain dosages were effective (see Fig. 20). The optimal hormone regimen was 40–80 micrograms EB with 250 micrograms once a week for four weeks. Large doses of EB (160 μg/week) did not facilitate marking under any progesterone regime. Owen also points out that only high-scoring females were behaviorally affected by gonadectomy. Previous studies may have failed to detect estrogen effects on female scent marking because they used randomly selected females, thereby including many low-markers, and large doses of EB (160 μg/week).

High-marking females also differ from low-markers in their ability to respond to hormone stimulation (Yahr and Thiessen 1975). Scent marking and ventral gland size were reinstated in ovariectomized high-markers by weekly injections of 40 micrograms EB or 40 micrograms EB followed by 250 micrograms P. The two hormone treatments were equally effective, but progesterone might have had synergistic effects if smaller EB doses had been used. We attempted to test this hypothesis (Yahr

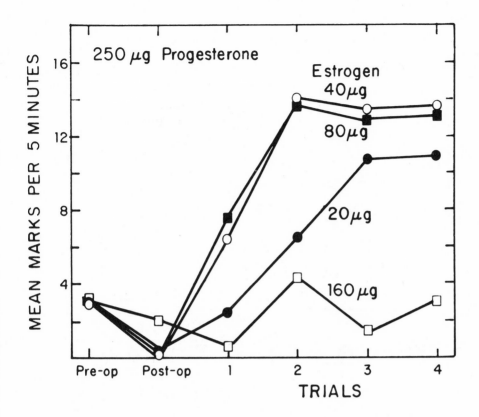

Figure 20.
Effects of ovariectomy and injection of various doses of EB (combined with 250 μg P) on territorial marking in females.

and Thiessen 1975), but too few high-marking females were available. Instead, we compared females that scent marked spontaneously with those that did not mark for their responses to low doses of EB. The spontaneous scent markers began marking again when given either EB (6–10 μg every 6 days) or EB plus P (400 μg). The two steroids did not interact to affect the behavior in spontaneous scent markers, and zero-markers did not react to either treatment. In other studies of females, Yahr (in press *b*) noted that a few zero-markers do begin marking at high levels when treated with EB or EB and P, but most of them do not react at all. TP seems more effective

than EB in this regard, but most zero-marking females are also insensitive to this androgen. As is the case for males, low-marking female gerbils have smaller scent glands than high-markers do. Yet, their glands attain the size of high-markers' glands after EB or TP therapy even though these hormones may not affect their behavior.

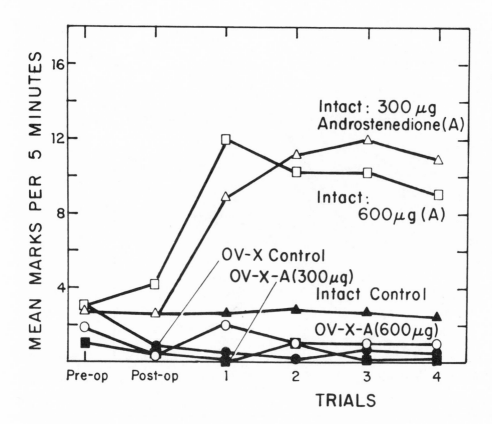

Figure 21.
Effects of androstenedione injections on territorial marking in ovariectomized and intact females.

Here, then, the picture of hormone control for the female gains some clarity. Low levels of marking, and marking which occurs prepubertally, do not appear to depend upon gonadal secretions. The basic neurological mechanisms may be encoded without hormones, but their full functioning does appear to rely on gonadal products, namely, estrogen or possibly estrogen plus progesterone. Testosterone can activate marking in the female, and estrogen but not progesterone stimulates marking in males. In other words, both sexes have steroid sensors for both sex classes of hormones, but they normally respond to their homotypical hormones—estrogen in the case of the female and testosterone in the case of the male.

Another of Owen's (1972) observations is noteworthy. Ovariectomized females did not respond to androstenedione (300–600 μg per week) with increases in marking, but intact females did (see Fig. 21). Androstenedione is an immediate precursor for testosterone. Probably gonadal tissue can convert it to testosterone and thereby induce marking. The next section gives evidence for a similar conversion in the male. The point still remains: estrogen and estrogen plus progesterone are the primary hormones activating high marking in the female.

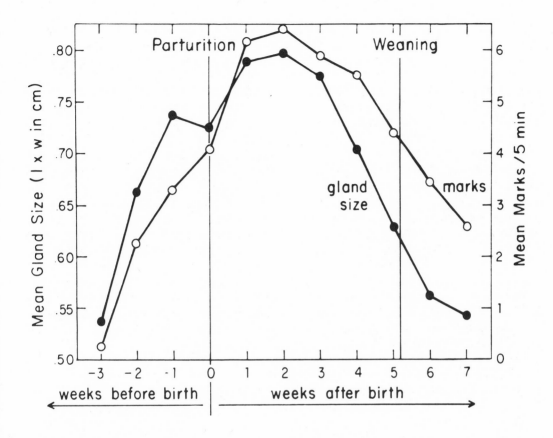

Figure 22.
Temporal pattern of marking behavior and gland size through the female reproductive cycle.

It is during pregnancy and lactation that the female displays her highest rate of scent marking (Wallace, Owen, and Thiessen 1973). In the standard testing area, increases in marking were apparent during the third trimester of gestation and reached a peak during the second week of lactation, several days prior to the weaning of the offspring. Following weaning and the cessation of lactation, ventral marking decreased to normal levels. The increase in marking during the reproductive period is quite phenomenal, sometimes reaching three to four times base levels, and it closely parallels changes in scent gland size, as seen in Figure 22. The correlation coefficient between these two variables is 0.81. As with hormone-induced scent marking in females, the lactation-induced increase in scent marking is most pronounced in females that have scent marked before mating (Yahr 1976). In general, though, lactation stimulates more scent marking in zero-markers than exogenous hormone therapy does.

Quite clearly, ventral marking and secretory activity of the sebaceous gland are correlated with critical periods of the reproductive cycle. Associated with these changes are alterations in nest-building behavior (Wallace, Owen, and Thiessen 1973). When females were allowed twenty-four hours' access to cotton placed in the food hoppers of their cages, the amount pulled into the cages and used to build nests corresponded to variations in ventral marking and gland size; that is, nest-building activity increased during the third trimester of pregnancy, peaked sometime after parturition, and fell again before the pups were weaned. This function is seen in Figure 23. Earlier we suggested that lactating females use their sebum to mark and identify their pups. This seems to be another function related to maternal responsibilities and increased sebaceous secretion.

The nature of the hormone control of reproductive activities in females is not known at this time, but it probably involves estrogen and progesterone, the primary hormones of gestation and lactation. Ovarian secretions regulate the marking and gland changes, as ovariectomy attenuates them. Whether or not nest building and pup retrieval are under similar control is not answered, although the same hormone regulation is likely. Pilot studies suggest that prolactin is ineffective in inducing marking in females, and we have incidental data which indicate that pup retrieval does not occur in nonpregnant and nonlactating females.

Several times we have noted that lactating females are extremely aggressive. Recently Bell and Maier (personal communication, 1975) have noted the same thing, showing that lactating females defend their nest area vigorously from all intruders, except their mates. Apparently there are several interlaced functions that are stimulated by pregnancy and lactation, all of which serve to maintain a family unit and insure that the genotype will be perpetuated. Based on the evidence we have, it seems that the hormones of reproduction stimulate nest building and defense as well as chemocommunication between the mother and her pups. The interrelated functions are most obvious during the time the pups must be cared for. Estrogen and progesterone secreted from the ovaries probably act parsimoniously to guide all of these critical activities.

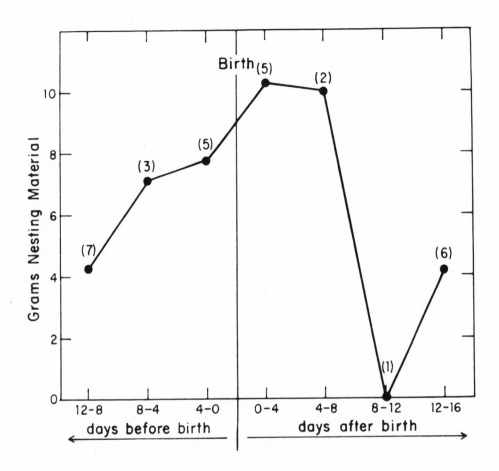

Figure 23.
Temporal pattern of nest-building activity through the female reproductive cycle.

5

Neurohormone Regulation of Territoriality in the Gerbil

Hormones interact with the genotype of an organism to regulate various behaviors, but the nature of this interaction is unknown. Clearly the genotype determines the character of the hormone and target tissue and sets the limits of behavioral modifiability. Yet almost nothing is known about the interaction between a hormone and the genome at the molecular level.

In most cases, the central nervous system is the primary target for hormone effects on behavior. Nevertheless, localization of neural function, whether hormonally regulated or not, is still sketchy for complex social processes. Territoriality and scent marking, in particular, have been explored very little. No one has specified sites in the nervous system of any species where these functions are integrated. Another persisting question in behavioral endocrinology is how a hormone affects the brain to activate a response. Few attempts have been made to describe the cellular processes that translate a hormone stimulus into a socially significant behavior. The closest approxima-. tion to these goals has been in the study of sex behavior. But even here the gaps in our knowledge are substantial (Beach 1970).

It is possible that hormone action within the central nervous system involves changes in gene action. Two separate areas of research suggest that hormones may regulate behavior by activating specific gene units in neural target tissues. First, there is substantial evidence that the elaboration of RNA and the synthesis of proteins are often specified by hormones (Clever 1966; Hamilton 1968; Tata 1966). At least in peripheral target tissues, it is becoming clear that many hormones stipulate the range of unique macromolecules that are produced. Second, there is increasing awareness that the dynamic quality of brain action involves genetic activity (Gaito 1971; Ungar 1970). This seems particularly apparent in learning processes. Considering these kinds of evidence, it is entirely plausible to suggest that hormones can induce a species-specific behavior as complex as territorial marking by activating unique DNA segments in localized brain areas.

Our initial probes into the molecular regulation of the territorial scent marking behavior of the gerbil are described in this chapter. We attempt to illustrate part of the fine-grained neurological control of territorial marking. A brain atlas of the gerbil is presented which provides denotative referents for neural localization and genetic studies. Evidence is presented for the critical role of the preoptic area in mediating hormone effects on marking; dose-response relationships are suggested for hormone implants into the hypothalamus; and the behavioral effects of a wide array of steroids are assessed in an attempt to define the molecular nature of hormones which elicit the response. Preliminary work is presented on our initial attempts to stimulate ventral marking with brain electrodes. Finally, summaries of studies which implicate gene activation and neurotransmitter functions are presented.

117

A Stereotaxic Brain Atlas of the Gerbil

The construction of a stereotaxic atlas for the gerbil brain has been essential for our work. Thiessen and Goar previously published an atlas of the hypothalamus (1970), and we have now completed an atlas of the entire brain (a similar atlas has recently been published by Loskota, Lomax, and Verity [1974]). Because it will be of general use to any investigator working with the gerbil brain, the atlas is presented in its entirety. It is based on adult males approximately 120 days of age.

Coronal sections 10 microns in thickness were projected as enlargements and traced in detail. The sections, 0.5 millimeters apart, range from 7.0 millimeters anterior to Bregma to 8.0 posterior to Bregma. The incisor and ear bars are horizontally mounted 34 and 30 millimeters, respectively, from the base plate. The plane of coronal sectioning at 8.3° is illustrated in Figure 24, and the dorsal and ventral positions of the sections are indicated in Figures 25 and 26. All coordinates were verified numerous times by implanting cannulae and sectioning the brain along cannulae tracks.

Table 20 gives abbreviations and common Latin names for all areas identified in our sections. Figures 27–57 present the drawings from our sections.

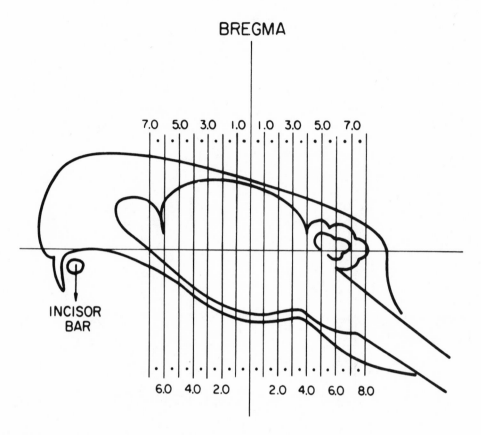

Figure 24.
Midsagittal section through gerbil brain illustrating sections in relation to head angle in stereo-taxic instrument.

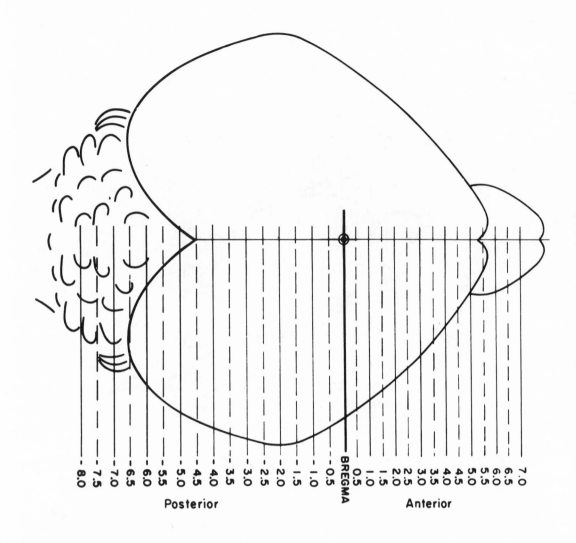

8.0 7.5 7.0 6.5 6.0 5.5 5.0 4.5 4.0 3.5 3.0 2.5 2.0 1.5 1.0 0.5 BREGMA 0.5 1.0 1.5 2.0 2.5 3.0 3.5 4.0 4.5 5.0 5.5 6.0 6.5 7.0

Posterior   Anterior

Figure 25.
Dorsal view of brain showing plane of sectioning.

120

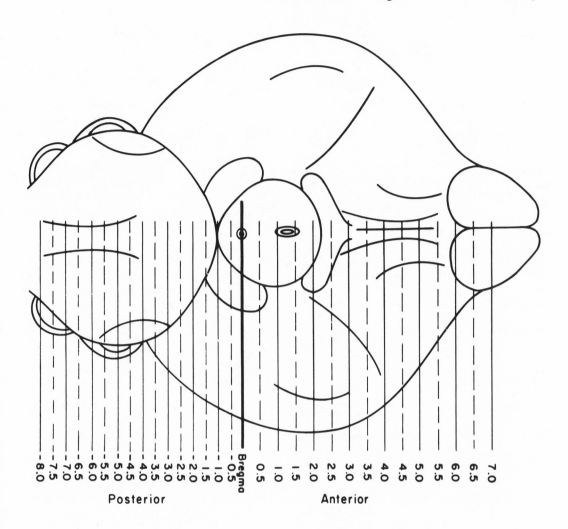

Figure 26.
Ventral view of brain showing plane of sectioning.

A	Aqueduct of Sylvius; Aquaeductus cerebri (Sylvii)		BCA	Nucleus proprius commissurae anterioris
AAA	Anterior amygdaloid area; Area amygdaloidea anterior		BCI	Brachium of the inferior colliculus; Brachium colliculi inferioris
ABL	Basal amygdaloid nucleus, lateral part; Nucleus amygdaloideus basalis, pars lateralis		BCS	Brachium of the superior colliculus; Brachium colliculi superioris
ABM	Basal amygdaloid nucleus, medial part; Nucleus amygdaloideus basalis, pars medialis		cI	Vermian lobule I; Lingula
			cII	Vermian lobule II; Ventral lobule of lobulus centralis
ACB	Lateral parolfactorial area; Area parolfactoria lateralis; Nucleus accumbens septi		cIII	Vermian lobule III; Dorsal lobule of lobulus centralis
			cIIIa	Sublobule IIIa
ACE	Central amygdaloid nucleus; Nucleus amygdaloideus centralis		cIIIb	Sublobule IIIb
			cIV	Vermian lobule IV; Ventral lobule of culmen
ACO	Cortical amygdaloid nucleus; Nucleus amygdaloideus corticalis		cV	Vermian lobule V; Dorsal lobule of culmen
AD	Anterodorsal nucleus of thalamus; Nucleus anterodorsalis thalami		CA	Anterior commissure; Commissura anterior
			CC	Corpus callosum
AHA	Anterior area of hypothalamus; Area anterior hypothalami		CCI	Commissure of inferior colliculus; Commissura colliculi inferioris
AL	Lateral amygdaloid nucleus; Nucleus lateralis amygdaloideus		CE	External capsule; Capsula externa
AM	Anteromedial nucleus of thalamus; Nucleus anteromedialis thalami		CF	Dorsal hippocampal commissure; Commissura fornicis dorsalis; Commissura hippocampi dorsalis
AME	Medial amygdaloid nucleus; Nucleus amygdaloideus medialis		CH	Hippocampal commissure; Commissura fornicis; Commissura hippocampi
ARH	Arcuate nucleus of hypothalamus; Nucleus arcuatus hypothalami		CI	Internal capsule; Capsula interna
			CIF	Inferior colliculus; Colliculus inferior
AV	Anteroventral nucleus of thalamus; Nucleus anteroventralis thalami		CL	Nucleus of Luys; Nucleus subthalamicus
			CLA	Claustrum

CO	Optic chiasma; Chiasma opticum		FI	Fimbria of hippocampus; Fimbria hippocampi
CP	Posterior commissure; Commissura posterior		FL	Fornix longus (of Forel)
CPU	Caudate nucleus; Putamen		fl.	Flocculus
cr.l	Crus I		FLD	Dorsal fasciculus of Schutz; Dorsal longitudinal bundle; Fasciculus longitudinalis dorsalis (Schutz)
CS	Superior colliculus; Colliculus superior			
CSC	Commissure of superior colliculus; Commissura colliculi superioris		FLM	Medial longitudinal bundle; Fasciculus longitudinal medialis
CT	Central tegmental nucleus; Nucleus centralis tegmenti		FP	Pyramidal fibers; Fibrae pyramidales
CTR	Trapezoid nucleus; Nucleus corporis trapezoidei		fpr.	Fissura prima
D	Nucleus of Darkschewitz		FR	Rhinal fissure; Fissura rhinalis
DBB	Diagonal band of Broca; Gyrus diagonalis		FVS	Fibrae vestibularis secondariae
DBC	Decussation of brachium conjunctivum; Decussatio brachii conjunctivi		FX	Fornix; Corpus columna
			GL	Lateral geniculate body; Corpus geniculatum laterale
DMH	Dorsomedial nucleus of hypothalamus; Nucleus dorsomedialis hypothalami		GM	Medial geniculate body; Corpus geniculatum mediale
			GP	Globus pallidus
DPCS	Decussation of superior cerebellar peduncles; Decussatio pedunculorum cerebellarium superiorum		H	Habenula
			HL	Lateral habenular nucleus; Nucleus habenularis lateralis
			HM	Medial habenular nucleus; Nucleus habenularis medialis
DTD	Decussation of Meynert; Decussatio tegmenti dorsalis		HP	Tractus habenulo-interpeduncularis; Fasciculus retroflexus (Meynert)
DTV	Decussation of Forel; Decussatio tegmenti ventralis			
DTZ	Trapezoid decussation; Decussatio corporis trapezoidei		HPC	Hippocampus; Cornu ammonis
EC	Pineal body; Epiphysis cerebri		ICL	Intercalated amygdaloid nucleus; Nucleus amygdaloideus intercalatus
ENT	Entorhinal cortex; Cortex entorhinalis			
FD	Dentate Gyrus; Gyrus dentatus; Fascia dentata		IP	Interpeduncular nucleus; Nucleus interpeduncularis
			LaCM	Lamina cellularum mitralium
FH	Hippocampal fissure; Fissura hippocampi		LaF	Lamina fibrosa
			LaG	Lamina glomerulosa

123

LaGl	Lamina granularis interna		NCV	Ventral cochlear nucleus; Nucleus cochlearis ventralis
LaM	Lamina medullaris			
LaPE	Lamina plexiformis externa		ND	Dentate nucleus; Nucleus dentatus
LaPI	Lamina plexiformis interna			
LC	Locus caeruleus		NI	Nucleus interpositus
LHA	Lateral hypothalamic area; Area lateralis hypothalami		NMT	Mesencephalic nucleus of trigeminal nerve; Nucleus tractus mesencephalici nervi trigemini
LL	Lateral lemniscus; Lemniscus lateralis			
LM	Medial lemniscus; Lemniscus medialis		NOT	Nucleus of olfactory tract; nucleus tractus olfactorii
LS	Lateral septal nucleus; Nucleus lateralis septi		NPT	Posterior nucleus of thalamus; Nucleus posterior thalami
l.sim.	Lobulus simplex		NPV	Medial vestibular nucleus; Nucleus principalis vestibularis
LT	Lateral nucleus of thalamus; Nucleus lateralis thalami			
LTP	Lateral nucleus of thalamus, posterior part; Nucleus lateralis thalami, pars posterior		NR	Red nucleus; Nucleus ruber
			NSV	Superior vestibular nucleus; Nucleus superioris vestibularis
MD	Dorsomedial nucleus of thalamus; Nucleus mediodorsalis thalami		NTST	Spinal nucleus of trigeminal nerve; Nucleus tractus spinalis nervi trigemini
MFB	Medial forebrain bundle; Fasciculus medialis telencephali		OA	Anterior olfactory nucleus; Nucleus olfactorius anterior
ML	Lateral mamillary nucleus; Nucleus mamillaris lateralis		OAD	Anterior olfactory nucleus, dorsal part; Nucleus olfactorius anterior, pars dorsalis
MM	Medial mamillary nucleus; Nucleus mamillaris medialis			
MP	Posterior mamillary nucleus; Nucleus mamillaris posterior		OAE	Anterior olfactory nucleus, external part; Nucleus olfactorius anterior, pars externa
MPA	Medial parolfactorial area; Gyrus parolfactorius medialis; Area parolfactoria medialis		OAL	Anterior olfactory nucleus, lateral part; Nucleus olfactorius anterior, pars lateralis
MPO	Medial preoptic area, Area praeoptica medialis		OAM	Anterior olfactory nucleus, medial part; Nucleus olfactorius anterior, pars medialis
MR	Nucleus medianus raphes			
MS	Medial septal nucleus; Nucleus medialis septi		OS	Superior olivary nucleus; Nucleus olivaris superior
MT	Mamillothalamic tract (bundle) of Vicq d'Azyr; Tractus mamillothalamicus (Vicq d'Azyr)		OT	Optic tract; Tractus opticus
			P	Pons
			PC	Cerebral peduncle; Pedunculus cerebri
NCP	Bed nucleus of the posterior commissure; Nucleus proprius commissurae posterioris		PCI	Inferior cerebellar peduncle; Pedunculus cerebellaris inferior

124

PCM	Middle cerebellar peduncle; Pedunculus cerebellaris medius	S	Subiculum
		SAM	Stratum album mediale colliculi superioris
PCS	Superior cerebellar peduncle; Pedunculus cerebellaris superior	SBV	Ventral spinocerebellar tract; Tractus spinocerebellaris ventralis
PF	Nucleus parafascicularis thalami	SC	Suprachiasmatic nucleus; Nucleus suprachiasmaticus
pfl.	Paraflocculus	SM	Stria medullaris thalami
PH	Posterior nucleus of hypothalamus; Nucleus posterior hypothalami	SN	Substantia nigra
		SNT	Nucleus of spinal tract of trigeminal nerve; Nucleus tractus spinalis nervi trigemini
PI	Internal preolivary nucleus; Nucleus praeolivaris internus	SO	Supraoptic nucleus of hypothalamus; Nucleus supraopticus hypothalami
PIR	Piriform cortex; Cortex piriformis		
PM	Mamillary peduncle; Pedunculus mamillaris	ST	Stria terminalis; Taenia semicircularis
PMD	Dorsal premamillary nucleus; Nucleus praemamillaris dorsalis	SUM	Area submamillaris
		TCS	Corticospinal tract; Tractus corticospinalis
PMV	Ventral premamillary nucleus; Nucleus praemamillaris ventralis	TL	Lateral tegmental nucleus; Nucleus lateralis tegmenti
		TOI	Intermediate olfactory tract; Tractus olfactorius intermedius
POA	Lateral preoptic area; Area praeoptica lateralis		
PRT	Pretectal area; Area praetectalis	TOL	Lateral olfactory tract; Tractus olfactorius lateralis
PT	Paratenial nucleus of thalamus; Nucleus parataenialis thalami	TP	Tractus tuberculopiriformis
		TPO	Tegmental nucleus of pons; Nucleus tegmenti pontis
PV	Paraventricular nucleus of thalamus; Nucleus paraventricularis thalami	TS	Nucleus triangularis septi
		TST	Root of spinal tract of trigeminal nerve; Radix tractus spinalis nervi trigemini
PVG	Central gray substance; Substantia grisea centralis		
PVH	Paraventricular nucleus of hypothalamus; Nucleus paraventricularis hypothalami	TT	Mamillotegmental tract; Tractus mamillotegmentalis
		TUO	Olfactory tubercle; Tuberculum olfactorium
RE	Nucleus reuniens thalami	TZ	Trapezoid body; Corpus trapezoideus
RF	Reticular formation of mesencephalon; Formatio reticularis mesencephali	V	Ventricle; Ventriculus
		VA	Ventral nucleus of thalamus, anterior part; Nucleus ventralis thalami, pars anterior
RH	Rhomboid nucleus of thalamus; Nucleus rhomboideus thalami		

VCLL	Ventrocaudal nucleus of lateral lemnicus; Nucleus ventralis caudalis lemnisci lateralis	VTN	Tsai's ventral tegmental nucleus; Nucleus ventralis tegmenti (Tsai)
VD	Dorsal nucleus of thalamus; Nucleus dorsalis thalami	ZI	Zona incerta
		II	Optic nerve; Nervus opticus
VE	Ventral nucleus of thalamus; Nucleus ventralis thalami	III	Oculomotor nerve; Nucleus of oculomotor nerve; Nervus oculomotorius; Nucleus nervi oculomotorii
VM	Ventral nucleus of thalamus, medial part; Nucleus ventralis thalami, pars medialis	V	Mesencephalic nucleus of trigeminal nerve; Nucleus tractus mesencephalici nervi trigemini
VMH	Ventromedial nucleus of hypothalamus; Nucleus ventromedialis hypothalami		
VO	Olfactory ventricle; Ventriculus olfactorius	VII	Facial nerve; Nervus facialis
VT	Ventral tegmental nucleus; Nucleus ventralis tegmenti		

7.0 mm Anterior to Bregma

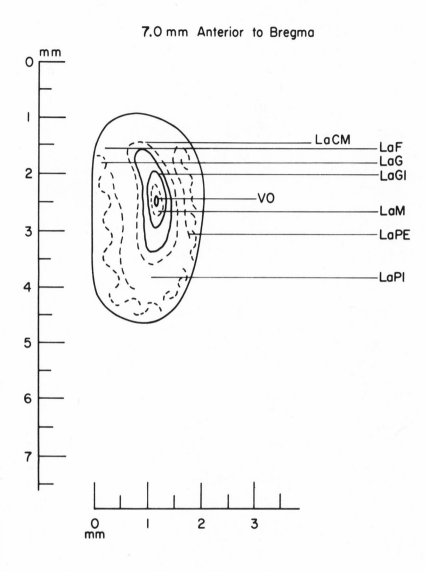

Figure 27.
Coronal Section 7.0 Millimeters Anterior to Bregma

6.5 mm Anterior to Bregma

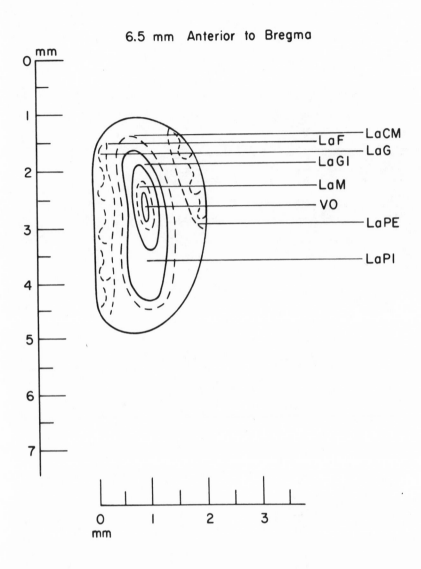

Figure 28.
Coronal Section 6.5 Millimeters Anterior to Bregma

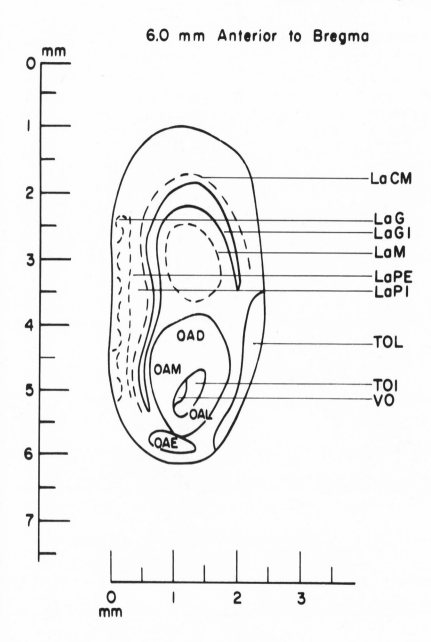

Figure 29.
Coronal Section 6.0 Millimeters Anterior to Bregma

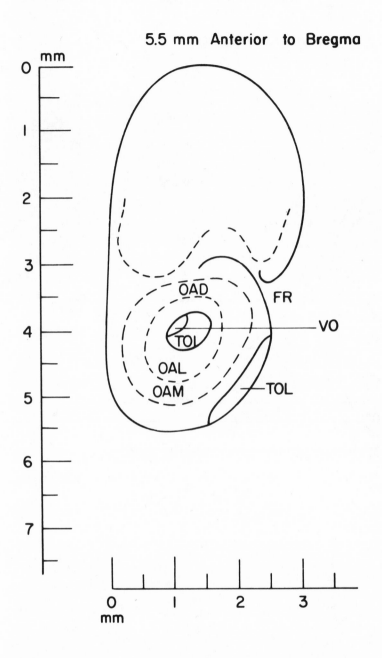

5.5 mm Anterior to Bregma

Figure 30.
Coronal Section 5.5 Millimeters Anterior to Bregma

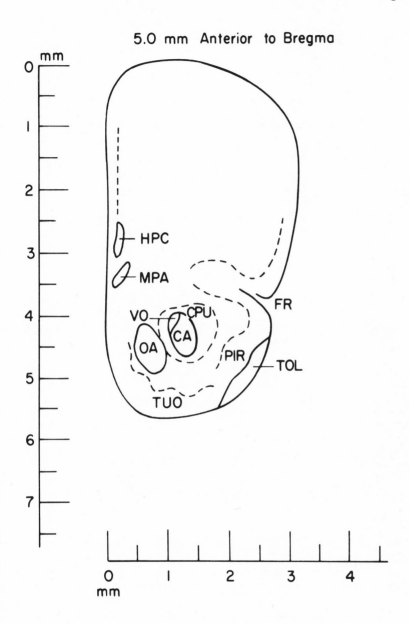

5.0 mm Anterior to Bregma

Figure 31.
Coronal Section 5.0 Millimeters Anterior to Bregma

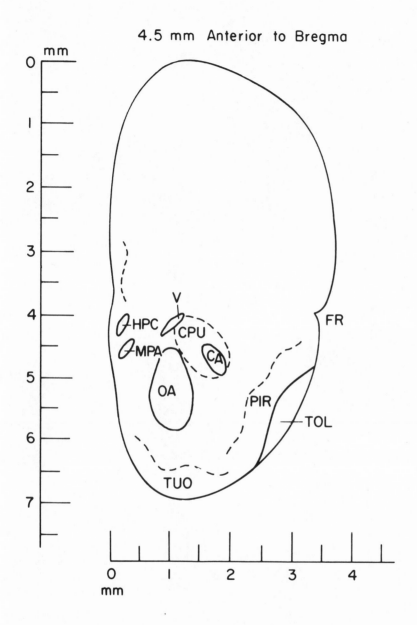

Figure 32.
Coronal Section 4.5 Millimeters Anterior to Bregma

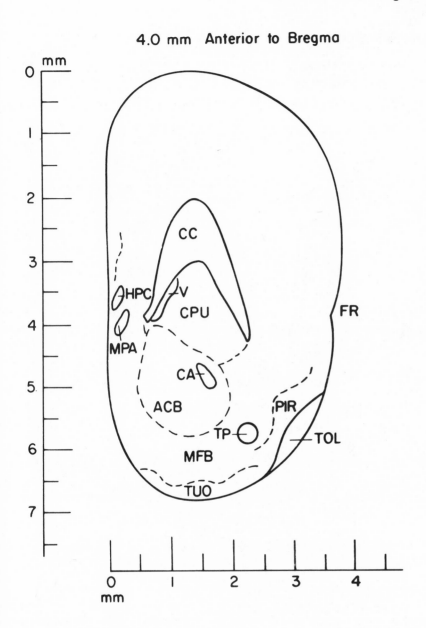

Figure 33.
Coronal Section 4.0 Millimeters Anterior to Bregma

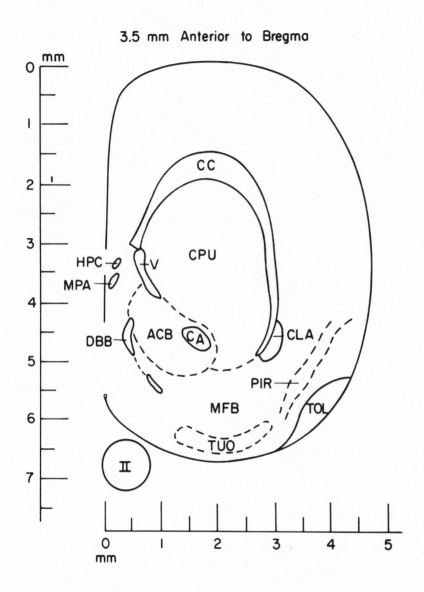

3.5 mm Anterior to Bregma

Figure 34.
Coronal Section 3.5 Millimeters Anterior to Bregma

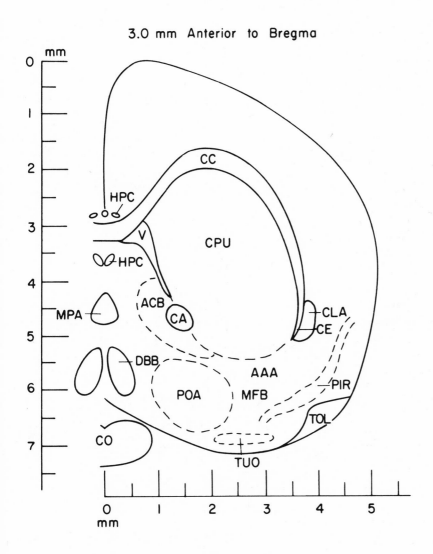

3.0 mm Anterior to Bregma

Figure 35.
Coronal Section 3.0 Millimeters Anterior to Bregma

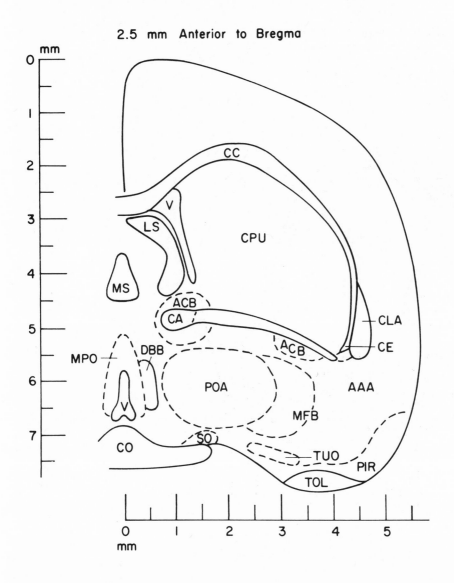

2.5 mm Anterior to Bregma

Figure 36.
Coronal Section 2.5 Millimeters Anterior to Bregma

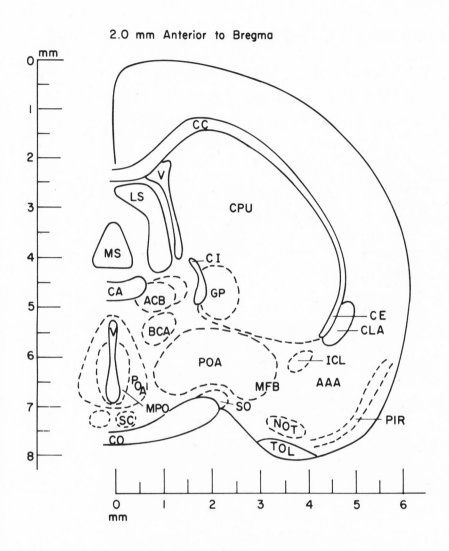

Figure 37.
Coronal Section 2.0 Millimeters Anterior to Bregma

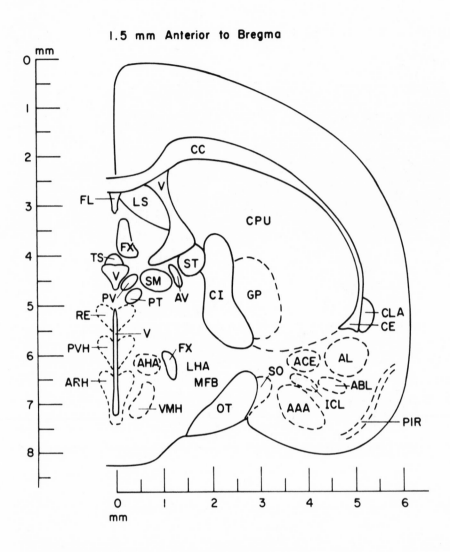

Figure 38.
Coronal Section 1.5 Millimeters Anterior to Bregma

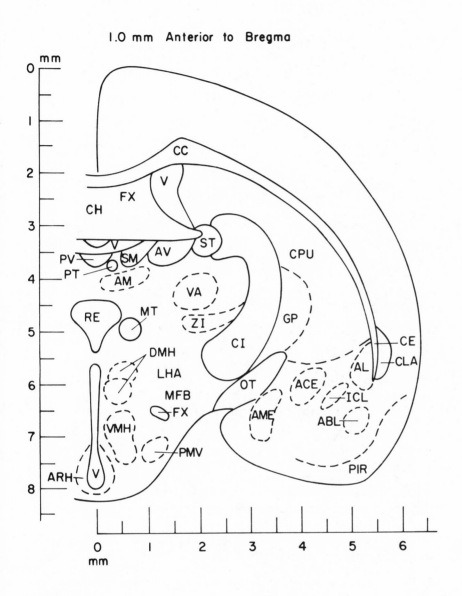

1.0 mm Anterior to Bregma

Figure 39.
Coronal Section 1.0 Millimeters Anterior to Bregma

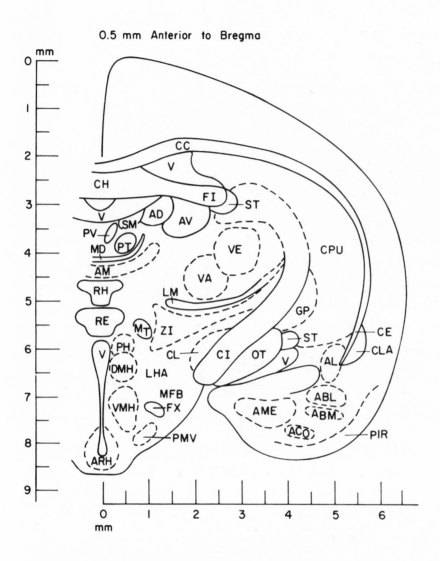

0.5 mm Anterior to Bregma

Figure 40.
Coronal Section 0.5 Millimeters Anterior to Bregma

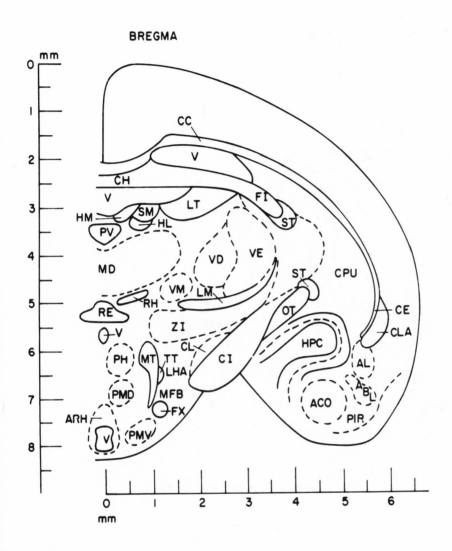

Figure 41.
Coronal Section of Bregma

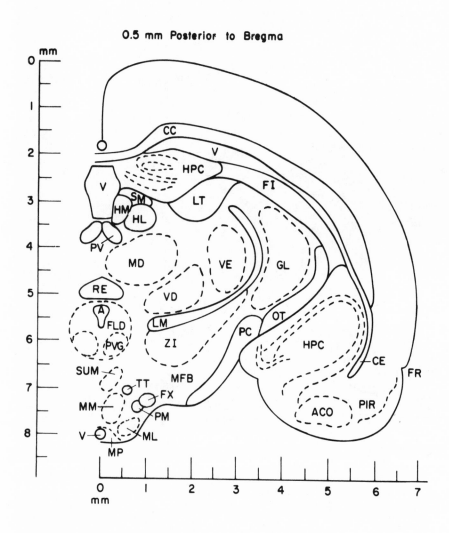

Figure 42.
Coronal Section 0.5 Millimeters Posterior to Bregma

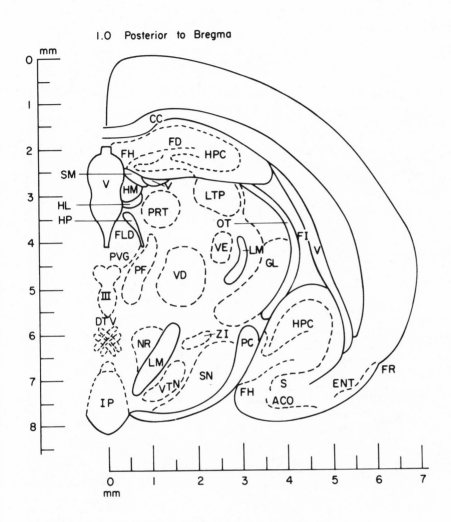

Figure 43.
Coronal Section 1.0 Millimeters Posterior to Bregma

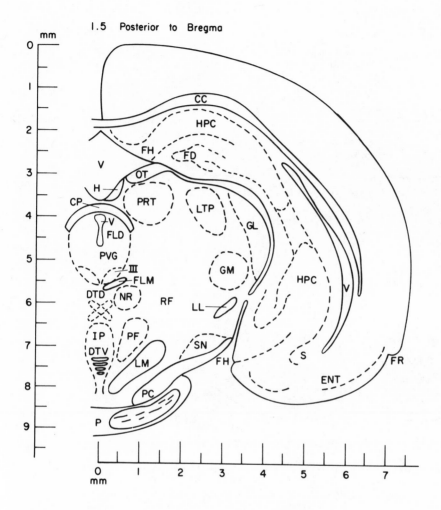

Figure 44.
Coronal Section 1.5 Millimeters Posterior to Bregma

144

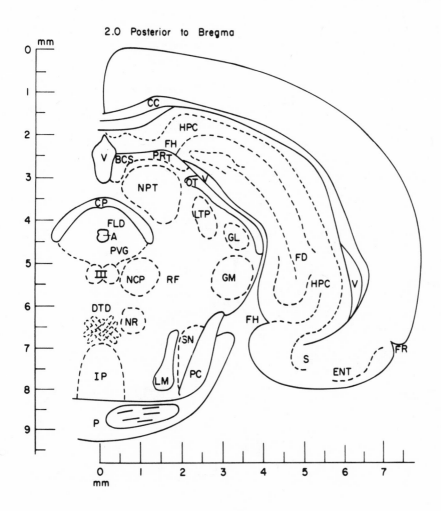

Figure 45.
Coronal Section 2.0 Millimeters Posterior to Bregma

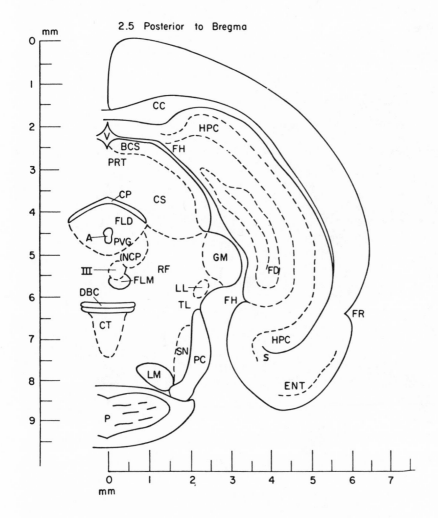

Figure 46.
Coronal Section 2.5 Millimeters Posterior to Bregma

146

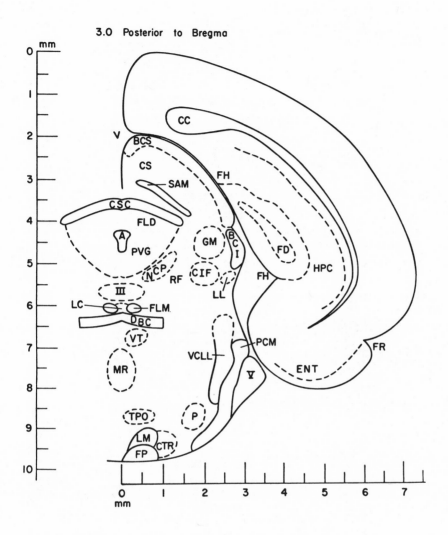

Figure 47.
Coronal Section 3.0 Millimeters Posterior to Bregma

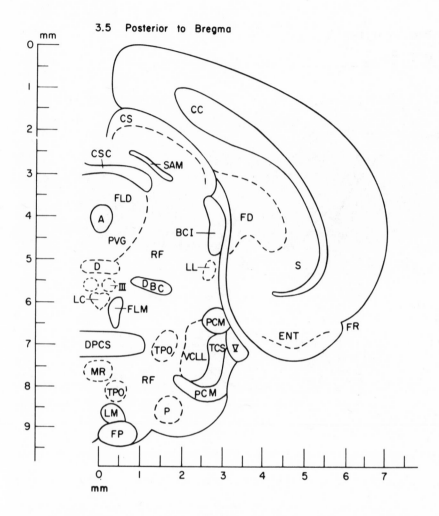

Figure 48.
Coronal Section 3.5 Millimeters Posterior to Bregma

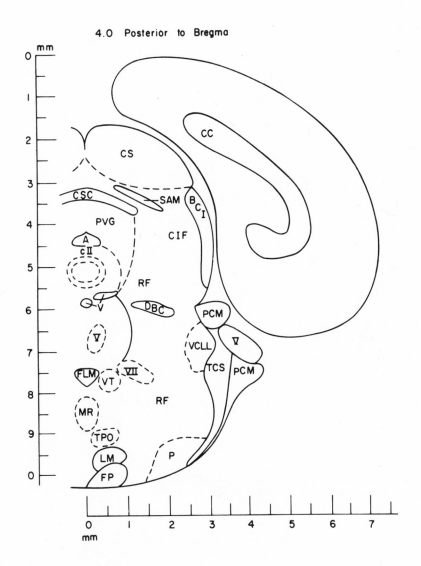

Figure 49.
Coronal Section 4.0 Millimeters Posterior to Bregma

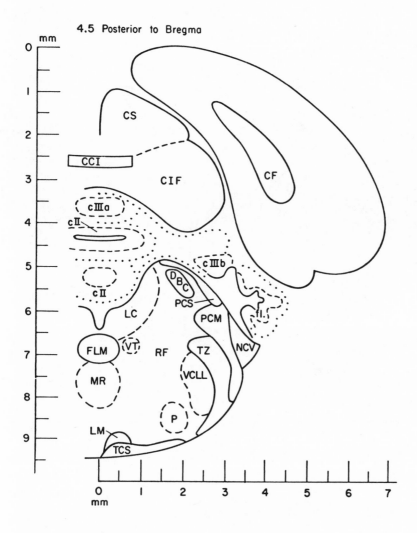

Figure 50.
Coronal Section 4.5 Millimeters Posterior to Bregma

150

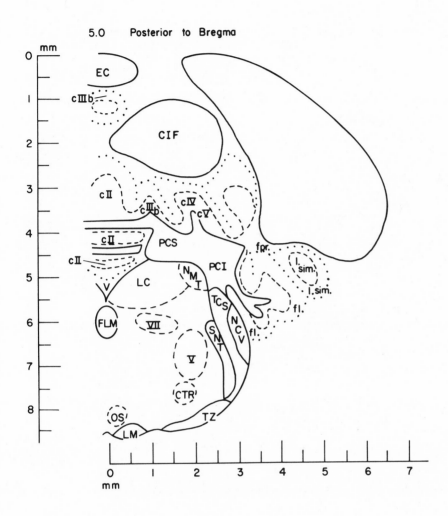

Figure 51.
Coronal Section 5.0 Millimeters Posterior to Bregma

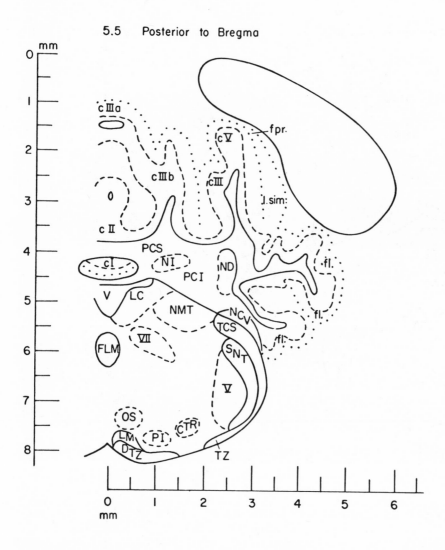

5.5   Posterior to Bregma

Figure 52.
Coronal Section 5.5 Millimeters Posterior to Bregma

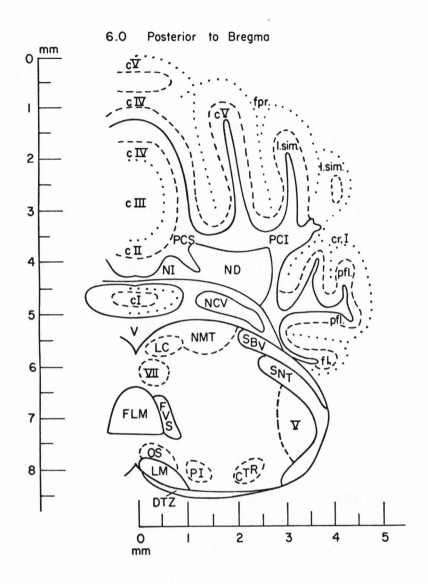

Figure 53.
Coronal Section 6.0 Millimeters Posterior to Bregma

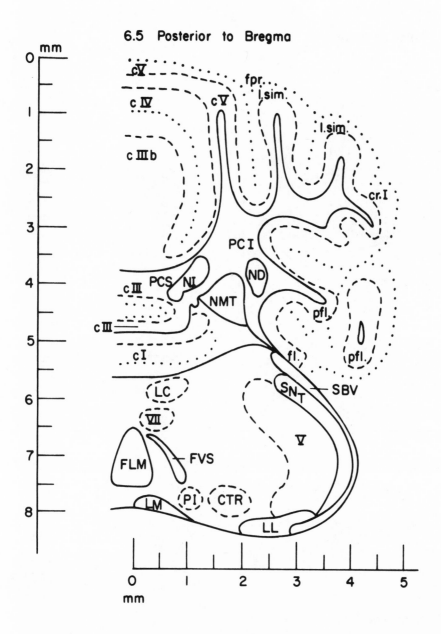

6.5 Posterior to Bregma

Figure 54.
Coronal Section 6.5 Millimeters Posterior to Bregma

154

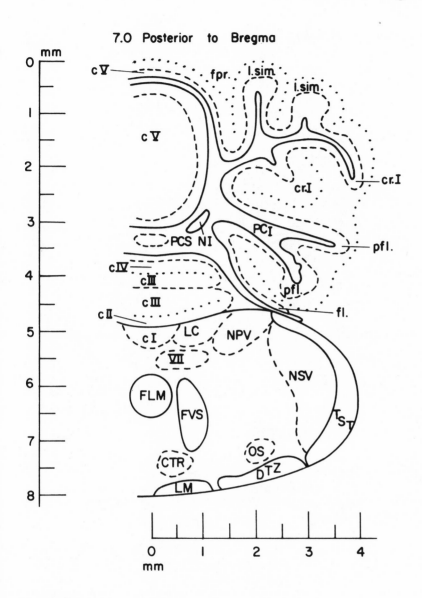

Figure 55.
Coronal Section 7.0 Millimeters Posterior to Bregma

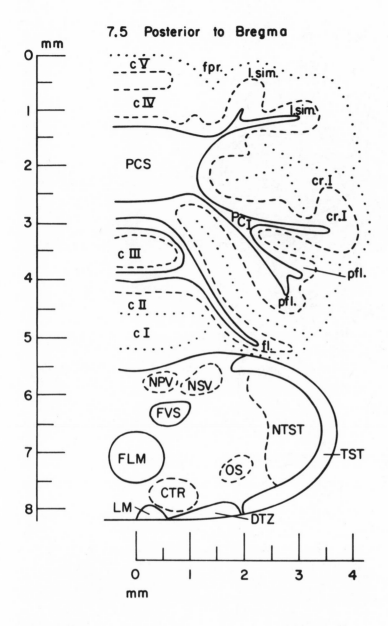

Figure 56.
Coronal Section 7.5 Millimeters Posterior to Bregma

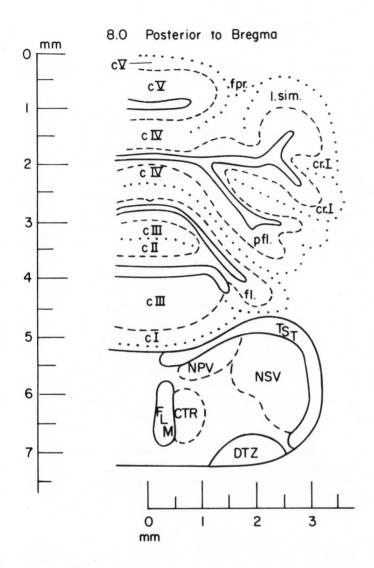

Figure 57.
Coronal Section 8.0 Millimeters Posterior to Bregma

Neural Localization of Hormone Control of Marking

Gonadal steroids regulate territorial marking in male and female gerbils. Gland size and sebum secretion are also determined by hormone titers but show only a gross correlation with the marking response. Marking responds more rapidly to hormone variation and can be pushed to supernormal levels, although gland size cannot. In fact, the gland is not even necessary for marking to occur (Blum and Thiessen 1970).

These facts led us to speculate that central mechanisms mediate territorial marking. We have since verified this hypothesis in several ways. For example, a preliminary study showed that TP applied directly to the brain induced marking in male castrates at very low doses (Thiessen and Yahr 1970). As little as 1.25–5.00 micrograms TP injected into the lateral ventricle facilitated marking, whereas a minimum of 32 micrograms TP weekly was needed to evoke the behavior systemically (Blum and Thiessen 1971). This supports the notion that the hormone acts centrally.

To locate the site of hormone control of marking in the brain, cannulae containing concentrated steroid were implanted at specific neural sites, thereby confining hormone stimulation to a localized group of cells. The hypothalamus was an obvious place to begin. In several other species, gonadal steroids regulate reproductive behaviors by acting on the hypothalamus, particularly the preoptic area. Examples include male and female sex behavior in rats (Davidson 1966; Lisk 1962), female sex behavior in cats and rabbits (Harris and Michael 1964; Palka and Sawyer 1966), male sex and courtship behavior in doves and domestic fowl (R. J. Barfield 1965; 1967) and egg incubation in doves (Komisaruk 1967). We therefore probed the hypothalamus and other parts of the brain with hormone implants and observed the amount of marking evoked.

Initially, three hypothalamic areas were explored: the posterior hypothalamus, the lateral hypothalamus, and the preoptic area (Thiessen and Yahr 1970). Castrated males received testosterone or cholesterol (an inert steroid) implants at these sites, or they received testosterone implants in the cortex or hippocampus. Only castrates with testosterone implants in the hypothalamus were induced to mark, and at one point their average marking score exceeded the precastration level. The responses of the individual hypothalamic sites could be distinguished to some degree. Preoptic-area hormone implants stimulated marking in all males in that group, and one male showed a large increase in marking the day after implantation. As the distance of the testosterone implant from the preoptic area increased, fewer males responded and the latency to respond increased. Thus, of the areas tested, testosterone was sufficient to induce marking reliably only when placed in the preoptic area. The increase in marking observed with hormone implants in other hypothalamic nuclei was probably due to diffusion of the steroid to this sensitive site.

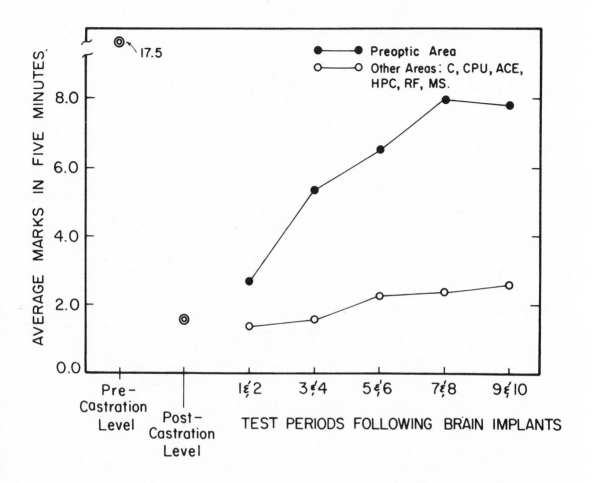

Figure 58.
Effects on territorial marking of castration and brain implants of TP. C = cortex; CPU = caudate nucleus; ACE = central amygdaloid nucleus; HPC = hippocampus; RF = reticular formation; MS = medial septal nucleus.

A related study examined androgen effects on various brain areas known to modulate hypothalamic activity or to be involved in olfactory processes (Thiessen, Yahr, and Owen 1973). These included the central amygdaloid, caudate, and medial septal nuclei, the hippocampus, the reticular formation, and the medial preoptic area (Feldman and Dafny 1970; Raisman 1970). The cortex was explored again as well. Only TP implants in the preoptic area reinstated reliable marking in castrates, as summarized in Figure 58. General activity, gland size, body weight, and seminal vesicle weight did not differ across groups, emphasizing that the behavioral effects of the hormone were specific and confined to the brain.

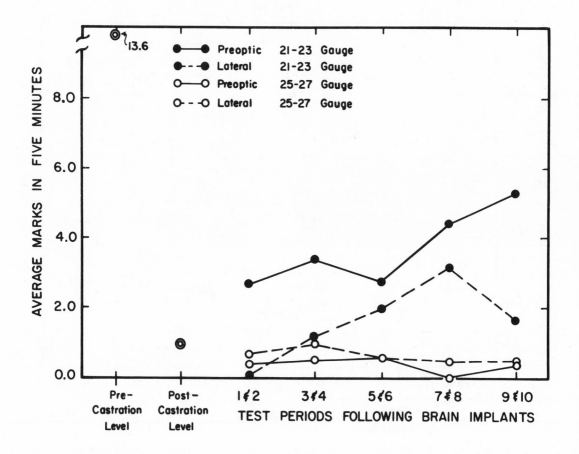

Figure 59.
Effects on territorial marking of castration and hypothalamic implants of TP. Large (21–23 gauge) and small (25–27 gauge) cannulae were implanted into the preoptic or lateral areas of the hypothalamus.

The gerbil's hypothalamic response to TP is also cannula-size (dose) dependent (Thiessen, Yahr, and Owen 1972). Small TP-filled cannulae (25–27 gauge) did not evoke marking in castrated males, whether implanted in the preoptic area or in the lateral hypothalamus. Larger cannulae (21–23 gauge) stimulated marking in both areas, although preoptic-area implants produced a larger, more rapid response (see Fig. 59). Once again, body weight and general activity differences were not involved, and ventral gland size correlated poorly with the marking response.

Steroid Restrictions on the Neural Stimulation of Marking

Only a few steroids elicit territorial marking; or, more precisely, preoptic area cells respond to only a limited range of steroid molecules. Our data suggest that the neural target tissue responds to a more restrictive set of steroids than do peripheral target organs. In other words, fewer steroids stimulate scent marking than stimulate gland growth and sebum secretion.

We have tested eleven steroids for their ability to stimulate marking in castrated males, both when administered systemically (500 micrograms twice a week) and

when implanted directly into the brain (Yahr and Thiessen 1972). The steroids used were androgens and their metabolic precursors, common ovarian steroids and precursors, and steroids which differed from testosterone at discrete points on the molecule. Most of the steroids fit into more than one of these categories. Figure 60 specifies the steroids that were used and also summarizes their systemic effects. Testosterone, estradiol-17β, estriol, and 19-nortestosterone were the only steroids that enhanced marking systemically. The

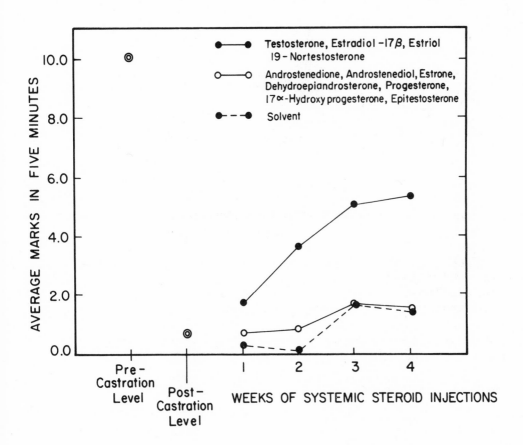

Figure 60.
Effects on territorial marking of castration and systemic injections of various steroids.

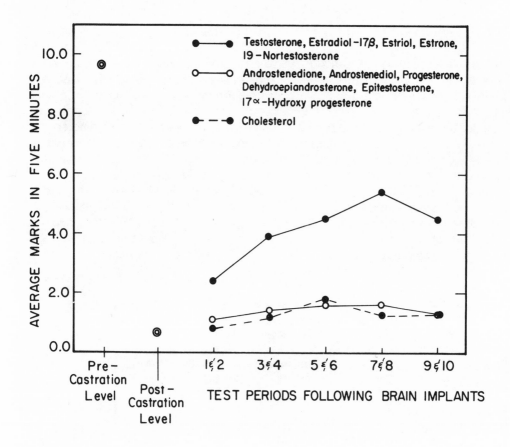

Figure 61.
Effects on territorial marking of castration and preoptic-area implants of various steroids.

same four steroids, plus estrone, were also effective when implanted in the preoptic area (see Fig. 61). Thus, of the steroids tested that are normally secreted in males in significant amounts (Eik-Nes 1970), only testosterone reinstated territorial marking. This restriction may be the result of selection pressure to coordinate territorial behavior with factors that regulate testosterone secretion, including age, stress, population density, season, and social status.

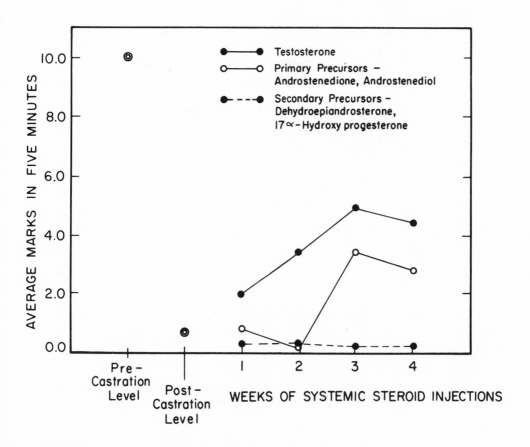

Figure 62.
Effects on territorial marking of castration and systemic injections of testosterone or its primary or secondary precursors.

The behavioral effects of testosterone precursors provide additional evidence that only testosterone regulates scent marking in males. The amount of marking that precursor steroids induced systemically depended on their position in the normal pathway of testosterone formation. Steroids which can be converted to testosterone in one metabolic step (androstenedione and androstenediol) induced more marking than steroids which require two steps for this conversion (dihydroepiandrosterone and 17 α-hydroxyprogesterone). This difference is shown in Figure 62. In contrast, neither primary nor second-

ary precursors elevated marking when implanted in the preoptic area, as shown in Figure 63. This suggests that the primary precursors stimulate scent marking only after their conversion to testosterone. Perhaps such conversions are more likely to occur systemically than centrally.

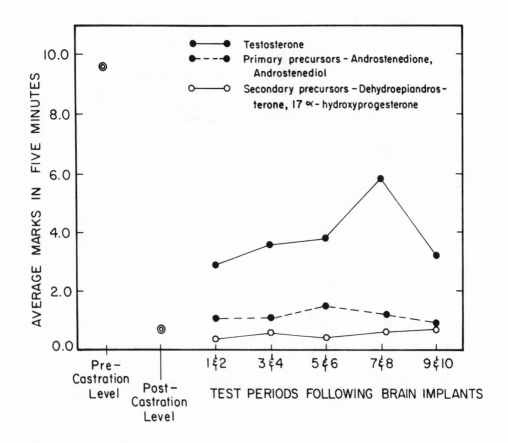

Figure 63.
Effects on territorial marking of castration and preoptic-area implants of testosterone or its primary or secondary precursors.

As mentioned earlier, the ventral scent gland responded to a wider range of steroids. Only two steroids, dehydroepi-androsterone and 17 α-hydroxyproges-terone, failed to cause gland growth. These same two steroids and epitestosterone failed to stimulate sebum secretion. In other words, the steroids which facilitated mark-ing behavior did not account for the major effects on gland function in either case.

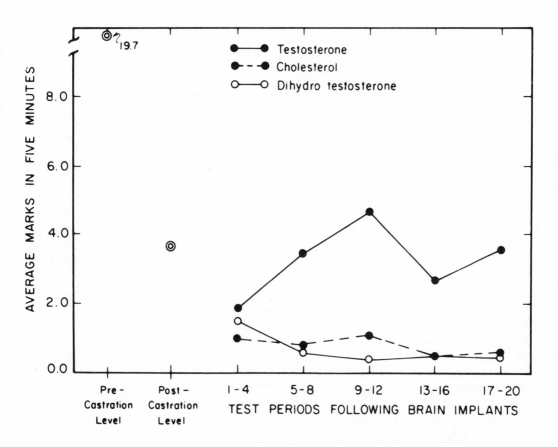

Figure 64.
Effects on territorial marking of castration and preoptic-area implants of TP or DHT.

The effects of another steroid, 5 α-dihy-drotestosterone (DHT), were examined independently (Thiessen, Yahr, and Owen 1973). DHT, a metabolite of testosterone, is considered to be the active form of the hormone molecule in peripheral target organs. Prostate cells, for example, accumulate testosterone and convert it to DHT, which is bound to chromatin in the cell nuclei and retained (Anderson and Liao 1968; Bruchovsky and Wilson 1968). To test the possibility that DHT mediates androgen effects on scent marking, we compared the marking scores of castrated males implanted with testosterone, DHT, or cholesterol in the preoptic area. As observed several times before, testosterone implants elevated marking scores. DHT, however, like cholesterol, had no behavioral effects (see Fig. 64), though it did cause ventral gland growth. Similar results were obtained by injecting DHT subcutaneously into randomly selected females; TP injections facilitated marking in such females, but injections of DHT did not.

In several other species, DHT has failed to mimic testosterone effects on the central nervous system. For example, DHT altered phallus sensitivity to TP but did not cause anovulatory sterility when given to infant female rats (Luttge and Whalen 1970). DHT did not promote mating in female rabbits or in castrated male rats (Beyer, McDonald, and Vidal 1970; McDonald et al. 1970; Whalen and Luttge 1971); yet each of these responses can be evoked with TP. To account for their data, McDonald et al. (1970) suggested that androgens, such as testosterone, which can be aromatized to form estrogens are more effective than non-aromatizable androgens, such as DHT, for stimulating androgen-dependent behaviors. The observations on gerbil scent marking were basically consistent with this aromatization hypothesis, although discrepancies were also apparent. For example, androstenedione, which is aromatized as efficiently as testosterone (Ryan 1960), does not elicit scent marking, whereas 19-nortestosterone, which is aromatized only 20 percent as efficiently as testosterone (Ryan 1960), induces high levels of marking.

An alternative hypothesis to account for the effects of different steroid hormones on gerbil scent marking was based on structural differences. The steroids that facilitated marking in male gerbils when administered systemically had two molecular features in common with testosterone (a 17 β-hydroxyl group and only one hydrogen at carbon 4) which did not appear together in any of the ineffective steroids. These features would distinguish testosterone from other steroids normally secreted in males and are the smallest number of characteristics which accurately do so. It was therefore speculated that central receptors recognized testosterone molecules by these key features (Yahr and Thiessen 1972).

To distinguish between this hypothesis and the one based on androgen aromatization, the effects of two more androgens, 1 α-methyltestosterone and 6 α-fluorotestosterone, were studied (Yahr in press a). Both of these androgens have the two structural features discussed above, but neither can be aromatized (Gual, Morato, Hayano, Gut, and Dorfman 1962). The 1 α-methyltestosterone did not reinstate scent marking in castrated male gerbils, but 6 α-fluorotestosterone propionate was as effective, dose for dose, as TP (see Fig. 65). Thus, the data did not support either hypothesis of the mechanism of hormone reaction. The 6 α-fluorotestosterone propionate was also as effective as TP for increasing scent gland size, but it promoted only half as much growth of the seminal vesicles as TP did.

Steroid restrictions on the stimulation of marking are also evident in females. As noted in the previous section, testosterone, estrogen, and estrogen plus progesterone activated scent marking in some ovariectomized females when given in appropriate systemic doses. Testosterone and estrogen were also effective in combination (600 μg TP plus 160 μg EB), but other steroids given systemically to females did not elicit marking (Owen 1972). Estrone, progesterone, pregnenalone, 17 α-hydroxyprogesterone and androstenedione were equally ineffective in gonadectomized females.

When various steroids were implanted centrally in females, only testosterone and estrogen stimulated marking (Owen 1972). Females selected for moderate to high marking scores were ovariectomized and implanted with TP, EB, TP plus P, EB plus P, TP plus EB, androstenedione, or cholesterol in the preoptic area. TP and EB reinstated the response, were about equally potent, and did not reveal antagonistic or additive actions when implanted together.

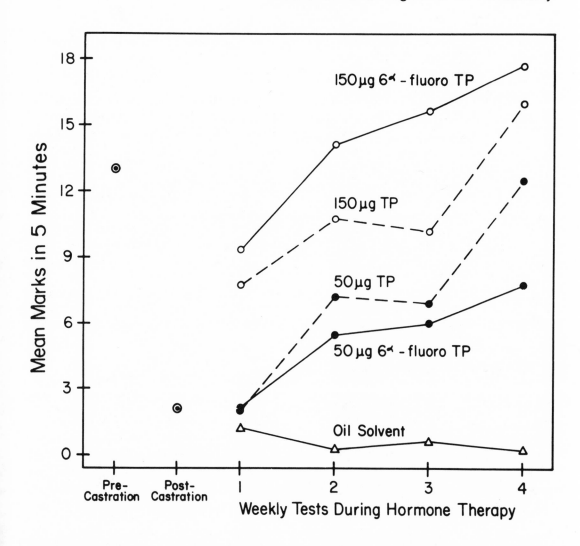

Figure 65.
Effects of testosterone propionate (TP) versus 6 α-fluorotestosterone propionate (6 α-fluoro TP), at two different dose levels, on the scent marking behavior of castrated male gerbils. The hormones were subcutaneously injected twice a week in the doses specified.

Progesterone did not enhance the effectiveness of either hormone. Androstenedione implants were ineffective in both females and males. Earlier, we mentioned that systemic injections of androstenedione stimulated marking in intact, but not ovariectomized, females. Considering both these results, it appears that the ovaries convert androstenedione to another steroid, possibly testosterone, which actually enhances the behavior.

As the results above suggest, the preoptic area is involved in hormone control of scent marking in females as well as in males. Using bilateral implants of EB dissolved in paraffin, Owen, Wallace, and Thiessen (1974) studied the responsiveness of various brain sites to estrogen stimulation in ovariectomized female gerbils. Over a four-week testing period, EB implants into the preoptic area, the septum, and the anterior hypothalamus induced significant increases in scent marking, whereas implants into the hippocampus, amygdala, thalamus, and anterior olfactory nucleus did not. The anterior hypothalamus and the preoptic area implants stimulated the most scent marking. On the last test, females with EB implants in the anterior hypothalamus marked more than each of the other groups, but on all of the other tests EB stimulation of the preoptic area and of the anterior hypothalamus was equally effective. Thus, the available data suggest that the neural sites involved in hormone control of scent marking are the same in both sexes of gerbils.

Electrical Evocation of Territorial Marking

The preoptic area may indeed be the most responsive site for steroids, but other brain areas probably contribute to the total effect. The relationships between the olfactory system and the hypothalamus preclude any simple mechanism. For example, reference to Figure 66 shows known connections between the olfactory bulb and other structures of the limbic system (Komisaruk and Beyer 1972; J. W. Scott and Leonard 1971; Raisman 1966; White 1965). Virtually any or all of these structures may exert control over territorial behavior.

Our initial explorations using electrical stimulation of brain sites in intact males bear this out. With the participation of Mark Dawber we have stimulated regions of the following sites: olfactory bulb, anterior olfactory nucleus, various amygdala sites, lateral preoptic area, median forebrain bundle, and reticular formation. A fifty-hertz symmetric biphasic square wave was generated by a Grass S88 stimulator and passed through a flexible cable to a unilaterally implanted stainless steel bipolar electrode. The animal was free-moving at all times in a modified open field studded with six marking pegs. In most cases animals were stimulated at 50, 100, 200 and 500 microamperes for intervals varying from 300 milliseconds up to 60 seconds. Stimulus parameters were chosen to fall within the median range used by other investigators.

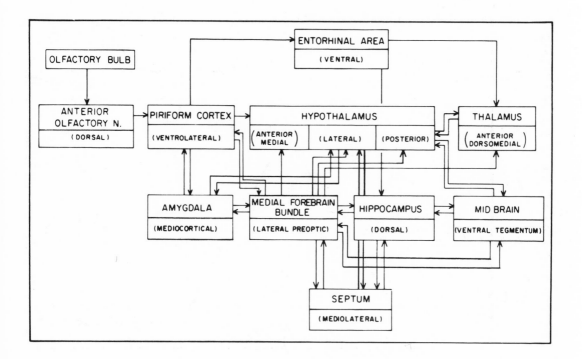

Figure 66.
Major olfactory and hypothalamic connections.

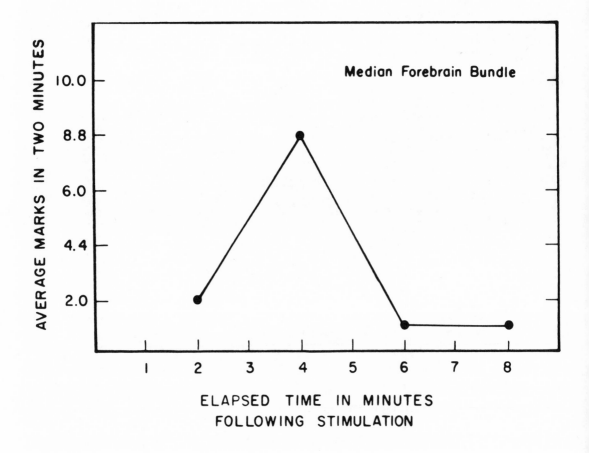

Figure 67.
Elevation of marking frequency following stimulation of the medial preoptic area.

Although our results are preliminary, it is clear that ventral marking can be evoked by stimulation in several areas, including the olfactory bulb, the dorsal anterior olfactory nucleus, the dorsal anterior and medial aspect of the medial forebrain bundle, and the anterior and basal aspects of the amygdala area. In total nine points of stimulation were effective in seven diverse areas. In nearly every case the effective stimulation parameters were fifty-hertz, one hundred microamperes, with a duration of 750 milliseconds. In all cases of effective stimulation there was a period of recruitment after the offset of the stimulus before marking occurred. The marking appeared in its full complexity and appeared normal in every way. A typical change in marking frequency following stimulation is shown in Figure 67.

Other sites are being explored and attempts to replicate our preliminary findings are underway. Areas for exploration include those shown in Figure 66. We are also varying testosterone titer in an attempt to determine which structures interact with the hormone to produce the response. It is possible that some nuclei can be provoked without testosterone present, whereas others may require the hormone priming at those or connected sites.

Gene Action and Territorial Marking

Relatively little is known about the mechanisms of hormone action on the brain in any species. In contrast, considerable progress has been made in understanding the mechanism of hormone action on peripheral target tissues. It is known, for example, that some hormones, such as thyroid hormone and gonadal steroids, stimulate RNA and protein synthesis in target organ cells (Hamilton 1968; O'Malley and Means 1974; Tata 1966). As a result of the hormone-induced changes in RNA and protein synthesis, the target organ cells perform different functions. For instance, when estrogen reaches the cells of a chick's oviduct, it causes a new form of RNA to be transcribed from the DNA in the nucleus (O'Malley and Means 1974). This new type of RNA codes, in turn, for ovalbumin, an egg yolk protein. Thus, the ability of estrogen to stimulate egg production in chickens results, at least in part, from its stipulating which proteins the oviduct cells produce.

We hypothesized that hormones affect preoptic area cells in a similar way; that is, hormones may alter neural metabolism in the preoptic area at the level of genetic transcription. Conceivably, hormone-induced changes in RNA synthesis, and subsequently in protein synthesis, could modify neurotransmitter levels (or turnover rates), which in turn may regulate territorial scent marking. We speculated that the hormone would increase the levels of an excitatory transmitter.

Various predictions of the gene action hypothesis have been tested, using drugs which disrupt the metabolic sequence from DNA via RNA to protein at specific steps in the chain. For example, the antibiotic actinomycin D halts the transcription of RNA from DNA by intercalation between guanine and cytosine residues on the DNA strand (Sobell, Jain, Sakore, and Nordman 1971). Puromycin, another antibiotic, causes polypeptides to separate prematurely from the RNA template, thereby blocking the translation of RNA into protein (Williamson and Schweet 1965).

A series of three experiments examined the role of RNA transcription in hormone induction of scent marking (Thiessen and Yahr 1970; Thiessen, Yahr, and Owen 1973). In each case, testosterone or testosterone plus a small amount of actinomycin D was administered centrally to castrated males. Initially, ventricular injections of TP were employed. Later studies used cannula implants of testosterone in the lateral hypothalamus or the preoptic area. We have been critically sensitive to the possibility that the antibiotics might be toxic and thereby affect marking indirectly. Hence, we conducted LD_{50} studies prior to this work and used dosages within a range that did not usually affect body weight or general activity. In addition, only animals that were obviously healthy were included in the analyses. Under these rigid conditions actinomycin D consistently inhibited the increase in marking normally associated with testosterone. In fact, the males that received the antibiotic did not stop marking completely; rather, their scores remained at the levels characteristically observed after castration. The effects of actinomycin D and/or testosterone implants in the preoptic area are illustrated in Figure 68. The data suggest that genetic transcription may be an essential step in hormone stimulation of scent marking.

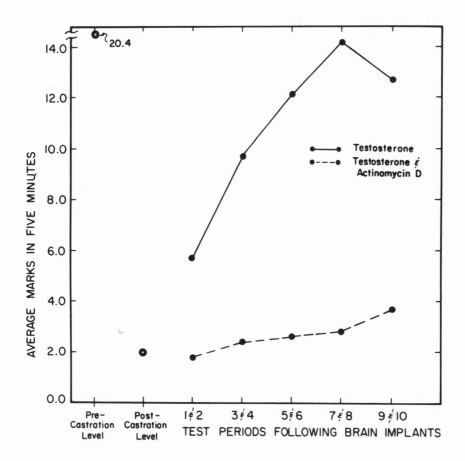

Figure 68.
Effects on territorial marking of castration and preoptic-area implants of testosterone or testosterone plus actinomycin D.

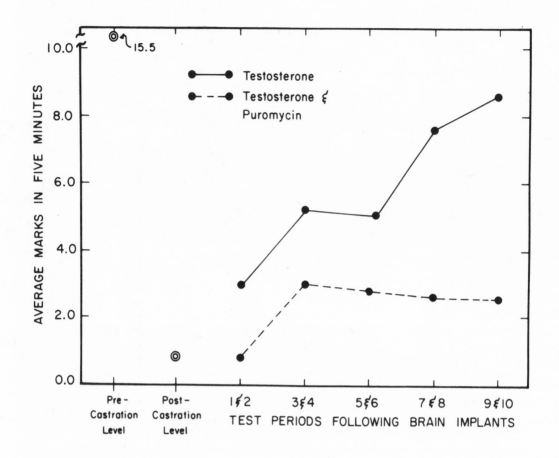

Figure 69.
Effects on territorial marking of castration and preoptic-area implants of TP or TP plus puro-mycin (first study).

Hormone effects on marking can also be blocked at the level of genetic translation. In two separate studies, castrated males received preoptic area implants of TP or TP plus puromycin (Thiessen, Yahr, and Owen 1972). In both cases, puromycin prevented the increase in marking normally produced by TP, although general activity was not affected by the drug. The effects of puromycin on marking are shown in Figures 69 and 70. These studies provide further support for the hypothesis that testosterone regulates territorial marking by altering gene activity in preoptic area cells.

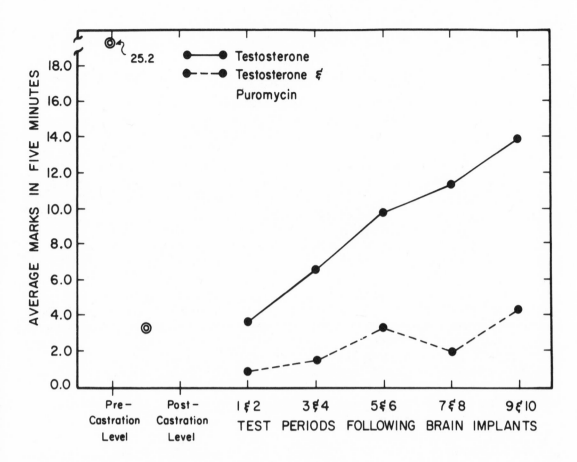

Figure 70.
Effects on territorial marking of castration and preoptic-area implants of TP or TP plus puromycin (second study).

RNase, a proteolytic enzyme that catabolizes RNA, also damped the behavioral response to TP (Thiessen, Yahr, and Owen 1972). Castrated males implanted with TP plus RNase consistently marked less than castrates implanted with TP alone; however, the repression was mild and short-lived. RNase is rapidly turned over in most cells, so the enzymes may have been inactivated before it could exert a significant effect.

We have also attempted to enhance hormone stimulation of scent marking by increasing RNA synthesis with magnesium pemoline, according to the procedure of Glasky and Simon (1966). Castrated male gerbils received TP, magnesium pemoline, TP plus magnesium pemoline, or cholesterol implants in the preoptic area. Magnesium pemoline increased marking slightly when given alone, but suppressed the marking response to TP. Possibly the drug and the hormone compete for the same receptor sites.

The pharmacological data support the notion that hormones regulate territorial marking by altering neural cellular metabolism, presumably through changes in gene activity in preoptic-area cells. As a further test of this hypothesis, we have looked for androgen-dependent changes in preoptic-area cellular components and have detected such changes, using immunochemical techniques (Yahr 1972). The rationale for using these techniques is simple. Assume that antibodies against five proteins from one tissue, A, are available. One way to determine whether another tissue, B, produces any of these five proteins is to test the reaction of the anti-A antibodies with the proteins from B. If, for example, three protein-antibody reactions occur, then we could conclude that three of the five proteins in tissue A also exist in tissue B, whereas two proteins in A are not present in B. The two tissues, A and B, may differ in regard to many other proteins as

well, and they may have many more proteins in common. In other words, this technique does not allow us to conclude that the tissues differ by only two proteins, but we can conclude that they differ by at least two.

In essence, then, this technique can be used to determine if two tissues produce different proteins (or other antigenic molecules), provided that antibodies against one or both tissues are available. We used this technique to determine if the preoptic area cells of castrated male gerbils contained different antigenic molecules (e.g., proteins) than the same tissue taken from males receiving hormone.

Briefly, the procedure was as follows. Four rabbits were immunized with preoptic-area tissue taken from TP-treated or oil-treated castrate male gerbils. Each type of tissue had been homogenized in buffered saline (Tyrode solution) and centrifuged; the supernatant and precipitate fractions were handled individually. Later, serum collected from each rabbit was tested for its reaction with the tissue fraction injected into that particular rabbit. This was simply to determine if antibody formation had occurred. Only the supernatant fractions had stimulated detectable antibody production. The two supernatant fractions were then tested with each of the two antisera by immunodiffusion and immunoelectrophoresis. With each technique three components were detected in preoptic-area tissue from oil-treated castrates, yet only two of the three components could be detected in preoptic-area tissue from castrates receiving TP. These results, diagrammed in Figure 71, suggest that testosterone suppressed the level of a particular component in preoptic-area cells. The nature of this component is unknown, but it could be related to neurotransmitter formation and the control of territorial marking.

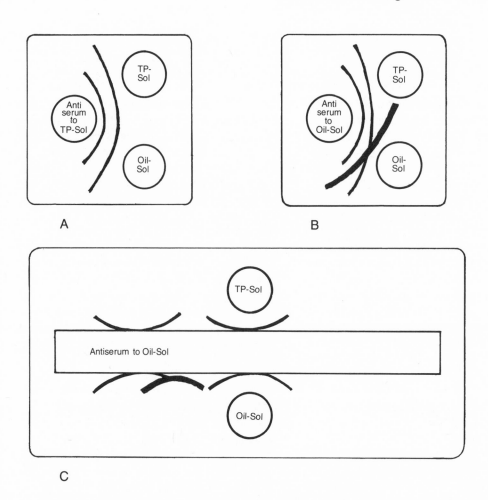

Figure 71.
Schematic diagram of the results obtained by two-dimensional immunodiffusion in an agar gel plate for: *A*. The reaction of Tyrode-soluble fractions of preoptic-area tissue homogenates from TP-treated castrates (TP-Sol) and oil-treated castrates (Oil-Sol) with antiserum to TP-Sol; and *B*. the reaction of TP-Sol and Oil-Sol with antiserum to Oil-Sol; also, *C*. the results obtained by immunoelectrophoresis in an agar gel plate for the reactions of TP-Sol and Oil-Sol with antiserum to Oil-Sol.

Changes in neurotransmitter activity within the preoptic area may mediate hormone induction of scent marking. Hence, a preliminary experiment (Yahr 1972) was performed to examine the interaction between androgen stimulation and variations in neurotransmitter levels in this critical area. Adult male castrates received preoptic-area implants of TP or TP plus a drug that alters the action of a specific neurotransmitter. Thus, the marking level induced by each combined preparation could be compared to that induced by the hormone alone. The effects of stimulating and inhibiting acetylcholine, catecholamines, and serotonin were individually examined in this way. The drugs used and their effects on territorial marking are listed in Table 21. In this study males implanted with TP plus atropine sulfate, an inhibitor of acetylcholine action, marked significantly more than males implanted with TP alone ($F = 4.88$; $p < .01$). Stimulation of acetylcholine action, using pilocarpine nitrate, suppressed TP effects on marking somewhat, and all the groups in which amines were affected marked less than the TP controls. There were no significant drug effects on general activity, ventral gland size, food or water intake, or body temperature.

The immunochemical data and the neurotransmitter effects are consistent with the notion that hormones alter neural cellular metabolism involved in territorial behavior. They also suggest that testosterone regulates territorial marking through *suppression* of an inhibitory transmitter, rather than by the induction of an excitatory transmitter as originally thought. Acetylcholine may play this inhibitory role. By blocking the production of an enzyme required to synthesize acetylcholine, the hormone may remove the inhibition and allow marking to occur.

Overall, the results suggest that the species-specific nature of territorial marking in the gerbil is genetically coded into cells of the preoptic area. Hormones tap this species-specificity and allow it full expression. It appears that the hormones (at least testosterone) act by stimulating the activity of some genes while inhibiting others. The end result of the hormone-gene interactions may be the modification of levels of neurotransmitter substances that regulate the final neuromuscular pathways leading to territorial behavior.

Table 21.
Summary of effects on territorial marking and activity of drugs
which alter neurotransmitter actions

Treatment (Drug Dosage)	Drug Action	N	Marks per Trial ($\bar{X} \pm$ SE)	Activity per Trial ($\bar{X} \pm$ SE)
TP + atropine sulfate (35 $\mu g/\mu l$ oil)	Inhibits acetylcholine action	8	5.5 ± 1.5	90.4 ± 8.5
TP + pilocarpine nitrate (700 $\mu g/\mu l$ oil)	Enhances action of acetylcholine	6	1.0 ± 0.6	106.6 ± 13.3
TP + phenoxybenzamine hydrochloride (350 $\mu g/\mu l$ oil)	Inhibits catecholamine action	7	0.8 ± 0.4	83.8 ± 9.0
TP + catechol (70 $\mu g/\mu l$ oil)	Enhances catecholamine action	8	1.9 ± 0.9	83.8 ± 4.9
TP + parachlorophenylalanine (700 $\mu g/\mu l$ oil)	Inhibits formation of serotonin	8	0.7 ± 0.1	86.3 ± 10.5
TP + 5-hydroxytryptophan + iproniazid phosphate (350 μg + 350 $\mu g/\mu l$ oil)	Enhances action of serotonin	6	1.8 ± 0.9	90.3 ± 7.5
TP + parachlorophenylalanine + catechol (700 μg + 70 $\mu g/\mu l$ oil)	Inhibits serotonin formation and enhances action of catecholamines	10	2.2 ± 1.0	84.0 ± 5.9
TP only	Control	8	2.8 ± 0.5	83.8 ± 6.4

Note: Each drug was mixed with safflower oil in the given doses. That mixture was then used to dampen crystalline TP to form a paste which was tamped into the cannulae.

6

The Harderian Gland: A New Pheromonal System

Our recent work indicates that an additional pheromone system is at work in the Mongolian gerbil and possibly other vertebrates. It involves a little-known structure, the Harderian gland. This pheromone complex may relate directly to ventral scent marking and operate under similar circumstances. Details of the work described here are taken from work by Thiessen, Goodwin, and Clancy (1976), Thiessen, Clancy, and Goodwin (1976), and unpublished observations by several colleagues (Jim Perkins, Mike Eddins, Mike Graham, and Sheila Marcks). Basic bibliographic material can be found in the first two mentioned papers.

The Harderian gland was first described by J. Harder in deer in 1694. The gland is wrapped behind the eyeball and is phylogenetically and anatomically associated with the nictitating membrane. It is prominent in rodents, rudimentary in higher primates, and totally absent in fish and aquatic amphibia. The morphology of the gland was described in various species from 1890 to 1925. Developmental and physiological regulation was explored without definitive results from 1925 to 1950. Since 1960 aspects of both the anatomy and physiology of the gland have been reinvestigated, using electron microscopy and improved histological techniques. Attention has focused on the presence of the pigment porphyrin, ranging from the biochemical definition of porphyrin and its isomers to the specifications of the genetic loci coding for porphyrin in Harderian tissue. Currently there are three primary and overlapping hypotheses concerning its biological function. The gland has been considered to act in immunological and bacterial defenses. It has been suggested that it provides a cushion for the eyeball and a lubricant for the eye. Finally, it has been thought of as an extraretinal photoreceptor in neonatal rodents, perhaps involving the mediation of the pineal gland. Some evidence exists to support each of these hypotheses; yet there is still no universal agreement as to its exact functions.

We now have evidence that the Harderian gland in the Mongolian gerbil (*Meriones unguiculatus*) and possibly in other vertebrates has a significant function in social behavior. It releases a pheromone which is actively spread over the face and body with the aid of saliva, thereby affecting the behavior of conspecifics. Our data support the notion that physiological arousal and thermogenesis lead to a simultaneous activation of the Harderian gland, release of saliva, and initiation of grooming. Grooming has the important functions of spreading saliva and Harder's pheromone and initiating attention and investigation by conspecifics. The Harderian material, in turn, evokes grooming in associates and leads to behavioral and physiological synchrony.

The Harderian Complex

Figure 72 depicts the Harderian gland *in situ* in the Mongolian gerbil. The gland is much larger than the eye itself and superficially resembles the spleen. A single gland from a male weighing approximately seventy grams measures 15.0 millimeters by 5.0 millimeters by 2.5 millimeters. The Harderian glands are larger in the male than in the female, but this difference largely disappears when corrected for body weight. Thus far we have seen no morphological or weight differences in Harderian glands between the sexes, but more detailed study may bring these to light. The gland is roughly hourglass shaped, wrapping over and around the optic nerve and filling the orbital cavity behind and under the eye. The gland is highly vascularized, heavily innervated, and densely pigmented. The gland is in fact a complex of smaller lobules encapsulated by a tough outer membrane. The smaller lobules are formed by myoepithelial cells ensheathing large secretory cells. These in turn are of two fundamentally distinct types, one of which produces lipids and the other, pigment. Ultrastructural analysis of other species reveals that neural and circulatory structures are closely juxtaposed. The primary innervation is from the inferior branches of the facial nerve, and the major vascularization is derived from the opthalmic artery.

Neural and possibly hormonal stimulation cause the gland to discharge its merocrine secretion into the nasal corner of the eye. The generation of the secretion is analogous to the excretion of milk from the mammary glands. The myoepithelial cells contract, and the secretion is transported to the anterior corners of the eyes through a specialized duct system associated with the nictitating membrane. More importantly for social interactions, the secretion is also

transported down the nasolacrimal canal (Harder-lacrimal canal), from which it exits at the extreme anterior aspect of the nares. In the rodent, at least, there are few other glands, if any, of this magnitude which exit to the outside of the body and which produce secretions so copiously.

The secretion itself is a complex mixture of lipids, proteins, and the pigment protoporphyrin. The lipid and pigment portions can be extracted with chloroform. The extract has a distinct floral odor which may contribute to its pheromone qualities. Since protoporphyrin fluoresces under long-wave (366 nm) ultraviolet irradiation in the Mongolian gerbil and several other species, the gland, its secretion, the Harder-lacrimal canal *in situ* and the supernatant of the extract fluoresce intensely.

Harderianectomy is accomplished by snipping the nictitating membrane of an anesthetized animal and extracting the gland with a forceps. Bilateral removal has no obvious effect on general activity, urination or defecation in an open field, depth perception (visual cliff), frequency and periodicity of grooming, or androgen-dependent indices. There is a nonsignificant trend toward smaller body size, and the general color of the pelage is considerably lightened. This latter effect is due to a general reduction of lipids and pigment of the hair. For purposes of the experiments to be described, glandless animals are compared with sham-operated and normal controls. When the lipids and pigments are extracted it is done with three milliliters of chloroform per gland pair. The secretory status of an individual is determined by irradiating the body surface with long-wave UV (ultraviolet) and rating the intensity of the fluorescence. The new secretion is primarily found in the nostrils and nose, the

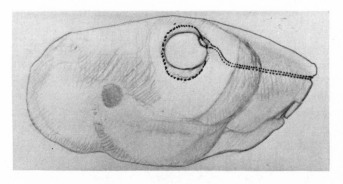

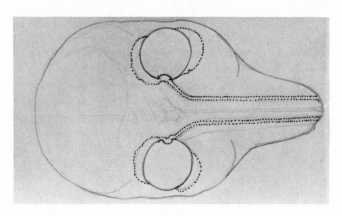

Figure 72.
The Harderian gland and Harder-lacrimal canal leading to the nares of the nose (*left*) and the areas of the face and paws (dotted areas) where Harderian material is spread (*right*).

chin, and the paws. We rate the intensity from 0 to 3 for each of these three body areas and total the score. Thus an animal may have a secretory status (UV status) ranging from 0, indicating no secretion, to 9, indicating very copious secretion and spread of the material.

Facilitation of Harderian Secretion by Grooming

Using the UV rating described, a randomly sampled group of gerbils from our colony was screened for the presence or absence of facial fluorescence. Males and females showed no obvious differences, but there was a clear age effect for the 375 animals sampled. Infants one to fifteen days of age were without fluorescence. At fifteen to thirty days 24 percent showed the pigment. Juveniles were up to 36 percent, and the adults showed the fluorescence in 54 percent of the cases. Thus there is an ontogenetic change, with an onset around fifteen days, when the full sequence of grooming appears, the eyes open, and other critical functions commence. It is not known if this change involves secretory rate, pigment formation, or active spreading of the pigment around the face, although in the mouse the age effect is associated with Harderian enzyme production and the ontogeny of grooming.

In adult animals of both sexes there is a high correlation between UV intensity ratings for members of a pair. That is, if one member of a pair has a great deal of secretion, then the likelihood of the second having copious secretion is also high. This synchrony is due to the mutual stimulation of Harderian secretion and facial grooming, as described below. A similar synchrony has been observed in larger populations. In general, an increase in Harderian secretion from the nose is associated with arousal. Excitement, stress, exposure to novelty, wheel-running activity, and elevation in ambient and body temperature are usually accompanied by an increased presence of fluorescent material (see below). We have noticed that fumes from volatile substances in such things as collodian and d-amphetamine will elicit a copious flow of the secretion. Other experiments now in progress tend to emphasize the importance of body temperature and vasodilation.

Of forty-three males tested for ten minutes in an open field, 93 percent groomed at least once, 53 percent groomed twice, and only 2 percent groomed three times. The mean latency for the first groom was 4.42 ± 0.33 minutes, and that for the second groom was 7.92 ± 0.30 minutes. For animals grooming twice during a ten-minute period the interval between grooms was 3.50 minutes. During the light half of a photoperiodic cycle males groom earlier during a trial and, as a result, more frequently. This may reflect a functional difference in thermal control and could have implications for olfactory signaling.

To establish the relationship between grooming and the spread of Harderian secretion on the face and paws, males and females were checked for UV status, placed singly in a novel cage, and then rechecked for UV status a few minutes later. An attempt was made to check half the number of males and females immediately after a groom and the other half at an identical point in time without a groom. Thus UV status change was assessed for groomers and nongroomers of both sexes. No sex differences were apparent, even though males generally groom more frequently.

The majority of animals that groomed showed a substantial increase in Harderian secretion, whereas most animals that did not groom showed no increase or a decrease in secretion. There were a few instances where increases were found without grooming (8 percent) or where no change or a decrease occurred with grooming (16 percent). However, most of these cases can be explained by the facts that some animals show a minimal increase which is dissipated by the time the animal is observed under UV and that partial grooms not meeting our criteria can sometimes spread the material. The relationship between grooming and the spreading of Harderian secretions is thus very strong.

Our evidence underlines the significance of grooming in gerbils and adds the important observation that grooming distributes Harderian secretions around the face and other parts of the body. Under conditions of novelty, grooming occurs rhythmically; one groom can be expected within five minutes and a second within the subsequent five minutes. The behavior seems contagious and induces similar secretory conditions in cage mates and population members. The exact nature of the provoking stimuli is unknown, although grooms occur under conditions of general arousal when sympathetic responses would be expected and when thermoregulatory responses come into play.

Functional Significance of Harder's Pheromone

Early in our investigations we devised a number of tests to determine if animals attend to the Harderian gland extract. For example, when animals were placed individually in a rectangular box studded with small metal pegs, animals directed their activity, ventral marking, and sniffing around the area containing 0.3 milliliter of Harderian extract. A similar volume of liver-bile extract did not elicit these spatially oriented responses. These results led to a series of studies which confirmed the relationship between Harderian secretion, pheromonal attraction, and facial grooming. Grooming has proved to be the critical behavior regulating the pheromonal process.

Our first indication that social behavior was altered by Harderianectomy was when male operates, sham-operates, and controls housed as pairs were separated for ten minutes for open-field testing and then observed for two minutes following repairing. Typically, when cage mates are re-

joined, mutual investigation of the facial and anogenital areas is frequent and intense. The interaction rarely lasts beyond two minutes. For these groups significant differences appeared in the number of facial investigations (sniffs) and the intensity of this investigation. Glandless animals engaged in significantly fewer facial investigations and investigated each other less intensely. Apparently animals without Harderian secretion are deficient in cues initiating mutual investigation.

Agonistic encounters were studied between Harderianectomized or sham-operated males and intact males for a fifteen-minute period in a neutral cage. Ten pairs of each type were observed. Measures included grooming, facial, and anogenital investigations, overt aggression, shaking, scratching, digging in sawdust, and squeaking. A running account was constructed of these behaviors during the observation period. Of these measures, the initiation of facial investigations and overt

185

aggression were depressed in the Harderianectomized-intact pairs. Anogenital investigations appeared to be more intense and of longer duration for the operates but did not occur more frequently. In every pair where an intact met an operate animal, the intact animal was clearly dominant. Sham-operated males were equally likely to win or lose a bout. No other measure distinguished the groups.

It was apparent that the intact animals directed their investigations toward the faces of the opponents, whereas the operates fixated on the genital area. Thirty-seven of the forty animals groomed, indicating that the lack of Harderian secretion in the operates has no effect on the grooming pattern. An incidental observation which requires further confirmation was that when the operates were paired with females following the male-male interaction, all females sexually rejected advances by the male. This was not true for sham-operated males, regardless of the outcome of the fight. Harderianectomized males may be less attractive as potential mates. In this and the previous experiment the sham-operated animals behaved in a way intermediate between the intacts and controls. This consistency suggests that disrupting the orbital ducts by excising the nictitating membrane, as we did for both the operates and sham-operates, may interfere with Harderian secretion.

Additional observations indicate that glandless animals smeared with Harderian extract (in the same facial pattern as that observed with intact animals) elicit more facial investigations by intact conspecifics than the same animals smeared with a control substance (stopcock grease). Equally interesting, sniffing of the animal smeared with Harderian extract is followed by an increase in full head grooms and partial grooms (momentary swipes with one or two paws). For this experiment males were given ten minutes of individual experience in an open field (Phase 1) followed by an additional ten minutes in which two stock-restrained Harderianectomized males were placed in the field (Phase 2). Only the heads protruded from the stock, separated from each other by approximately three centimeters. For experimental test males, one of the two stock animals was smeared with Harderian extract, whereas the second was smeared with stopcock grease. Control animals were confronted with stock animals both of which were smeared with stopcock grease. A number of stock pairs were used, and the position of the two animals in the stock was systematically alternated.

During Phase 2, experimental and control animals showed high interest in the stock-restrained pair. A high proportion of their ventral marks occurred in the immediate vicinity of the pair. Experimental animals, in addition, showed significantly more facial investigations of the Harderian-smeared animal, whereas control animals showed no differential preference between the two animals. Moreover, males exposed to the Harderian-smeared animals a significant increase in both full head grooms and partial grooms when compared with Phase 1. Harderian secretion obviously stimulates grooming responses and facial investigations. These social investigations include intense facial sniffing and licking, suggesting both olfactory and gustatory stimulation from the Harderian material.

There is an interaction between the act of grooming, social investigation, and the longevity of the facial fluorescence. When two males are placed in a cage and allowed to freely interact, a groom is generally followed by an investigation by the cage mate. As indicated for twenty males in Figure 73 the frequency of investigation is high during the groom and the following ten seconds and then diminishes to control levels by about seventy seconds after the groom. Typically when one member of a

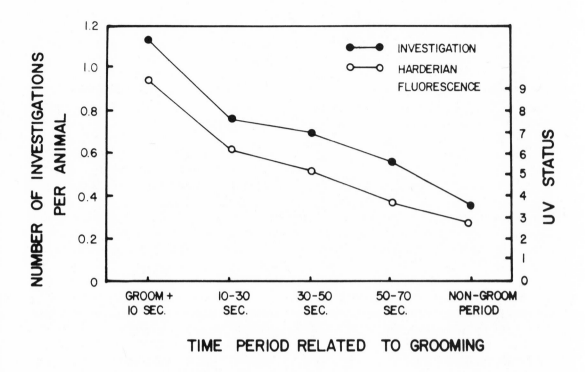

Figure 73.
Relationship between social investigation and intensity of facial fluorescence following a facial groom.

pair facial-groomed, the second stopped his activity, air-sniffed, and approached the groomer. Proximal investigations involved sniffing and licking around the nose, chin, and cheek areas. Grooming by one animal often appeared to stimulate grooming in the other. When the intensity of facial fluorescence was assessed in an additional eight males there was an obvious parallel between the intensity of investigation and the amount of fluorescence, suggesting that protoporphyrin intensity is an index of the active pheromone.

Grooming often oscillates between the nose and mouth, suggesting that Harderian and salivary secretions are mixed before being spread. Harderian pigment normally lasts approximately four minutes, whereas the extract continues to fluoresce for days.

When the extract is mixed with gerbil or human saliva the fluorescence dissipates in seconds. This demonstrates that saliva is not only used as a base for spreading Harderian material but may also subsequently deactivate the secretion. Denatured human saliva and gerbil saliva are without effect, suggesting an enzymatic reaction. While grooming is instrumental in the distribution of the secretion, grooming occurs in a rhythmic fashion even in glandless animals.

Both an olfactory and a gustatory cue could be important in groom-induced investigations. In one experiment six food-deprived males were trained to suppress lever pressing for food pellets when a Harderian extract was placed in an airstream. A lever press was followed by a brief foot shock when the Harderian material was present. Latency to bar press was strikingly increased during Harderian stimulation, indicating that the animals do indeed detect the pheromone by olfaction.

In another experiment adult male gerbils were tested for their ability to taste Harderian material, using a taste-aversion test. Essentially, animals (N = 6) were food-deprived, habituated to the daily presentation of twenty Noyes food pellets, with Harderian-smeared pellets followed immediately by intraperitoneal injections of semitoxic doses of lithium carbonate (1 percent body weight), and retested for ingestion of food pellets. Control animals (N = 8) were handled in a similar way except that they were injected with equal volumes of isotonic saline solution. Other animals were similarly treated with lithium carbonate, except that they either were made anosmic by flushing the nasal cavity with a 5 percent zinc sulfate solution (N = 8) or had the cavity flushed with an isotonic saline solution as a control (N = 8). Testing for taste aversion was conducted by measuring the consumption of both a Harderian-treated and a nontreated sample of Noyes pellets in two separate tests, with a single flavor at a time. Lithium-treated animals were reconditioned if they failed to show an aversion response to Harderian-smeared pellets.

Three of the six intact lithium-carbonate–injected animals showed Harderian aversion after one conditioning trial, two after two trials, and the last after a third trial. No saline-injected control animal showed an aversion. The lithium-carbonate and control groups differed significantly in consumption of Harderian-treated food but not plain food. Six of eight anosmic ($ZnSO_4$-treated) animals showed a specific aversion to Harderian-treated pellets. Similarly, all control (saline flush of nasal cavity plus Li_2CO_3) animals showed a significant aversion. Anosmia was verified following these tests by hiding a sunflower seed under the

sawdust of the animal's cage. Although all gerbils would readily eat when presented with sunflower seeds, only two of the eight zinc-sulfate-treated animals found the seed in ten minutes, whereas all the control subjects did. Apparently animals can react to Harderian material by taste, even when olfactory capabilities are eliminated.

Finally, it was of importance to demonstrate that olfactory cues from Harderian material could direct an animal's investigation even in the absence of another animal. Twelve adult male gerbils were tested singly for displayed interest in Harderian material during a fifteen-minute test in an open field. The Harderian extract, prepared as before, was rubbed on a cotton swab and hidden beneath a treadle in the floor. A second treadle on the opposite side of the box hid a control swab that had been treated in the same way but without Harderian material. The position of the stimulus was randomized between treadles. Each treadle tripped a separate timer, allowing a determination of the animal's time near the Harderian or the control swab. Time spent on one or the other side of the box was also recorded. In addition, the number of sniffs directed at each treadle was recorded, as well as the number of full facial grooms and partial grooms (momentary head swipes) on each side of the box.

Time and groom measures proved statistically unreliable because of large individual differences. The number of sniffs directed toward the treadle covering the Harderian material was significantly greater than the number of sniffs directed toward the opposing treadle. Apparently olfactory cues can elicit investigation independent of the presence of an animal and without gustatory cues.

The data as a whole indicate that grooming is an important aspect of social interactions and is responsible for the spread of an attractant pheromone from the Harderian gland. Olfactory cues stimulate a conspecific to investigate and lick the areas of the face where the Harderian material is spread. Recent work in our laboratory indicates that grooming is not accompanied by auditory or ultrasonic emissions, and we have not been able to demonstrate a visual signal with the fluorescent material. It appears that the initial reaction is based on olfactory molecules and is supplemented by gustatory cues. The exact function of the material is unknown, although there is a sex and dominance difference suggesting an important role during reproductive activities. Moreover, it could act to synchronize social and physiological activities and could be used in individual and status cues.

Thermoregulatory Control of Grooming and Harderian Secretion

The point has been made several times in earlier sections that increases in ambient or body temperature evoke facial grooming and saliva spreading in several species of rodents. This is no less true for the Mongolian gerbil. Preliminary data indicate a strong correlation between increases in ambient temperature, body temperature, and grooming. The typical pattern is for the animal to display full facial grooms until the body temperature approaches 37–38°C. Further increases in body temperature cause a switch in grooming pattern from the full groom to a partial groom which more effectively moves the saliva from the mouth to the ventral aspects of the face and body. Saliva is ordinarily placed on barred areas of the face which are extremely heat labile, but as the temperature increases copious amounts cover the entire ventral surface of the animal. Harderian secretion is equally intense under these conditions.

Other ways of increasing body temperature also activate facial grooming and Harderian secretion. For example, animals allowed to run in an activity wheel for three minutes quickly groom when they exit from the wheel. The eating of food pellets by food-deprived animals also hastens the onset of grooming. Finally, various doses of d-amphetamine stimulate saliva spreading and Harderian emission. All these manipulations are known to elevate body temperature (Cabanac 1974; 1975). Interestingly, amphetamine, an adrenergic mimic, releases Harderian flow minutes before salivation is activated; whereas pilocarpine, a cholinergic stimulant, causes copious salivation but no Harderian secretion. This latter finding suggests that both adrenergic and cholinergic systems are involved but that Harderian emission is controlled primarily by adrenergic fibers.

Amazingly, when animals are allowed to locomote over a thermal gradient, ranging in temperature from approximately 10° to 50° C, they choose to groom and ventral mark at approximately 29° C. Otherwise, the animals move rapidly over the gradient and show no obvious temperature preference. Interestingly, the temperature of the interface between the ventral aspect of the body and the nest is also around 29° C, suggesting that animals optimize their functions at this temperature. On the thermal gradient socially naïve, dominant, and subordinate males select this temperature to emit pheromones when tested individually, but when dominant and subordinate males are paired only the dominant animal grooms at this temperature. The subordinate animal is forced out of this zone and its grooming repressed. The data support the notion that external as well as internal temperature variations are critical in pheromone utilization.

Whatever the ultimate control, the evidence as a whole indicates that any stimulus which elevates body temperature will trigger a groom, salivation, and Harderian emission. Novelty, social interactions, and general activity are the prepotent stimulants within social organizations that can have these effects. While the primary adaptions associated with grooming are salivation and evaporative cooling, the intimate association between social interactions and thermoregulation has made it possible to superimpose a pheromone response during these critical periods. When animals evolved a terrestrial and endothermic existence the groom and salivation became critically important. At this time the Harderian gland became prominent and lent a secondary function to grooming that related to social persuasion and coordination.

Conclusions

Harder's pheromone, as we prefer to call it, appears to be a generalized primer which presets the conditions necessary for more specific social interactions. It acts to mutually arouse pairs or populations and to prepare them for other activities. Without this arousal stimulus, animals are relegated to a secondary status and either are ignored or assume an inferior social position. Based on our initial observations it appears that males failing to groom would not reproduce even in the absence of competition from other males. Temperature arousal and grooming are the keys for understanding the function of Harder's pheromone. Social excitement, or excitement in general, induces grooming and the spread of the pheromone. We suspect that vasodilation of the Harder-lacrimal tract, along with increased body temperature and peripheral stimulation of the ventral meatus by the paws during the grooming act, releases the pheromone. Grooming occurs independently of the presence of the Harderian gland, but grooming nearly always causes the spread of the secretion in intact animals, regardless of social and environmental conditions. This argues strongly that grooming is triggered by central events and that the spreading of the pheromone is evoked by peripheral stimulation, perhaps involving local dilation of the duct system in the nostrils. Of course Harderian secretion may also be evoked by direct neural activation of the gland itself. We have no evidence on this point. In any case, central and peripheral processes could be coadapted to produce an overall reaction, yet not be linked in a causal fashion.

One of the most interesting aspects of our observations is that animals synchronize their behaviors and pheromone production. Excitement induces temperature increases, grooming, and pheromone spreading, which in turn induce the same reaction in conspecifics. The animals, in effect, become physiologically and behaviorally entrained. While there are stimulus-response sequences imbedded in this interaction, the more important conclusion is that the animals reach an identical state of preparedness to act on any additional information. Without this preparedness other messages cannot be used adaptively.

The fate of the signal is probably dependent on enzymes in the saliva. The intensity of social investigation diminishes approximately two to three minutes following grooming. There is an intriguing coincidence between the sending time of the message, the half-life of the fluorescence of the secretion, and the interval between grooms. Later components of the groom are directed at the larger body areas as well as the face. This spreads the secretion more widely and diminishes the intensity of the fluorescence. The secretion may, when spread widely, function as a long-term body odor, denoting species or population identity.

We are not discounting other possible functions of the gland. It is ideally situated and programed to serve vision and eye and body lubrication as well as behavior. Indeed, its intimate anatomical association with the olfactory passage suggests a relation to respiratory and vomeronasal functions. The notion that the Harderian pheromone primes individuals and groups for the processing of more specific messages underlines the importance of studying signaling functions in combination. A medley of pheromones, plus input from other sensory systems, no doubt act in concert to integrate social activities. In our estimation, looking at integrated processes is the next profitable step in pheromone research.

We have no idea how general the Harderian pheromone is, or if its effects are species specific. We have seen similar relations between grooming and the spread of secretion by Sprague-Dawley (Holtzman) albino rats, golden hamsters (*Mesocricetus auratus*), white-tailed rats (*Mystromys albicaudatus*), and other species of gerbils (*Meriones tristrami*, *M. libycus*, and *M. shawi*). The relationships for the guinea pig (*Cavia porcellus*) and mouse

(*Mus musculus*) are unclear because external fluorescence is not always apparent. Secretion does occur, however. Eisenberg and Kleiman (1972) have seen eye secretions in tenrecs and suspect they are used for marking objects. These secretions could be from the Harderian gland. Deer, antelope, and other species often have orbital and preorbital secretions that are used for object marking. These may be Harderian in origin. The widespread appearance of Harderian glands in many endothermic species and the nearly ubiquitous behaviors of facial grooming, body investigation, and nuzzling suggest that Harder's pheromone may be a general communication signal among a wide range of vertebrate species.

7

Models for Territoriality

Man's interest in the territorial behavior of animals can be traced to Aristotle. Yet only recently have students of behavior attempted to reveal the physiological mechanisms underlying territoriality and to specify its evolutionary role. We feel that studies of Mongolian gerbils offer insights into the biology of territoriality and that this species is becoming a prototype for study. Its discrete scent marking response serves as a useful quantitative index of the physiological and social factors that alter territorial behavior. By no means can we answer the general question "Why do animals compete for space and scent mark territories?" but studies of gerbils are providing many tantalizing clues. At least for this species, we have information about the hormonal and neurological control of scent marking, and we are making strides toward understanding the social significance of the marking response.

In many respects, hormone control of gerbil scent marking seems homologous to hormone control of sex behavior and aggression in other mammalian species (see Table 22). Like sex behavior and aggression, territorial marking is essential for regulating populations and for perpetuating the species. For gerbils, territorial defense and scent marking may guarantee access to resources needed for reproduction, though scent marking surely serves other functions as well.

The complex sequence of neural and metabolic changes involved in hormone control of gerbil territorial marking has already been described in detail. Sensitive brain cells must detect steroid molecules and selectively take them up. Ultimately, this molecular stimulation at the cellular level must be translated into changes in neurotransmitter activity at the synaptic level. In gerbils, such a hormone receptor system apparently exists in the preoptic area of both males and females. Our data suggest that the steroid hormones act on the gerbil's preoptic area in generally the same way that they act on peripheral target tissues, that is, by altering genetic expression. More specifically, the hormones seem to stimulate the transcription of new RNA species or to increase the production of common ones.

The model in Figure 74, which has guided our research on this question, incorporates most of the findings discussed in Chapter 5 and projects the direction of our future research. The model is derived from others designed to explain general mechanisms of genetic transcription and translation (Britten and Davidson 1969; Hamilton 1968; Tomkins, Gelehrter, Granner, Martin, Samuels, and Thompson 1969) but can apply to behavior as well. As such, it maps out a relatively new strategy for the investigation of hormone-induced species-specific behaviors. It has enabled us to devise specific experimental hypotheses regarding hormone effects on scent marking and to test them systematically in the laboratory. More specificity is presented in the model than our completed experiments support, though our results are basically consistent with it.

Table 22.
Similarities between territorial marking by Mongolian gerbils
and the sexual behavior of rats

Behavioral Adaptations

The behavior occurs in a predictable sequence—exploration, approach to the stimulus, the primary act, and movement away from the stimulus.

Individual differences are large and some individuals do not display the behavior.

The behavior occurs only after habituation to the environment.

The behavior is sexually dimorphic.

Hormone Adaptations

The behavior develops at puberty. TP facilitates development of the behavior in males.

Performance of the behavior decreases gradually after castration in males and returns gradually after TP replacement therapy.

The first test before castration and the first test after hormone therapy begins yield highly variable scores.

Individual differences observed before castration reappear after hormone replacement.

The male's behavior can continue, at low levels, for a long time after castration.

Gonadal steroid hormones activate the behavior in both sexes.

Both sexes respond to the heterotypical steroid hormones.

DHT does not affect the male's behavior, although the secondary sex characteristics respond.

Androstenedione activates the behavior.

Reinstatement of the behavior in males after castration depends on the dose of TP used for replacement. Large TP doses induce rapid changes in behavior.

The behavior is more sensitive to androgen than the secondary sex characteristics are (scent gland and seminal vesicles in male gerbils and rats, respectively).

Expression of the behavior in males involves androgen-dependent structures (scent gland and genital papillae for gerbils and rats respectively), but these structures do not mediate androgen control of the response.

The male's behavior can be stimulated by TP implants into the preoptic area. Electrical stimulation of this and lower brain centers also activates the response.

The behavior is correlated with aggression and ultimately with gene transmission.

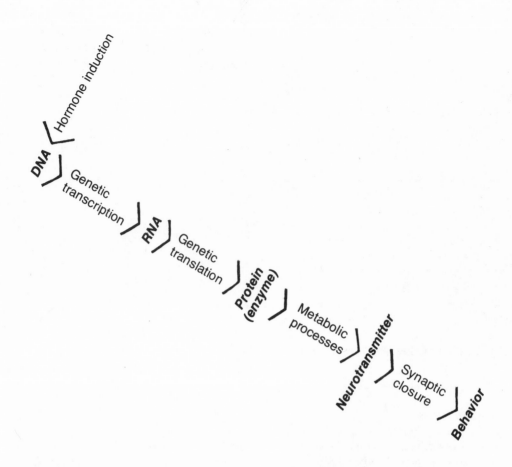

Figure 74.
Presumed relations between hormone induction of gene activity and behavior.

For example, the model implies that first RNA and then protein synthesis are necessary for androgen stimulation of marking in male gerbils. The suppressive effects of actinomycin D and puromycin on testosterone-induced marking uphold this prediction. These drugs block genetic transcription and translation, respectively. The decreased effectiveness of testosterone in the presence of RNase agrees with the model, and so do the enhancing and competitive effects of the RNA stimulant, magnesium pemoline. The failure of dihydrotestosterone to evoke marking in castrated males suggests that testosterone may affect preoptic-area cells directly, without being converted to a different "active form." An actual qualitative or quantitative change in protein synthesis has not yet been demonstrated, but the immunochemical data suggest that detectable changes in cellular components may be associated with steroid stimulation of marking behavior. The available data, however, point to testosterone repression of some cellular factor, rather than to new or increased production of macromolecules. Both processes may be involved, of course.

We have only begun to explore the possible mediating effects of neurotransmitters in this hormone-behavior relationship. The model in Figure 74 suggests that hormone-induced increases in an excitatory neurotransmitter are responsible for increases in scent marking. Presumably, though, the hormone could suppress the levels or turnover rate of an inhibitory transmitter, thereby removing inhibition and allowing scent

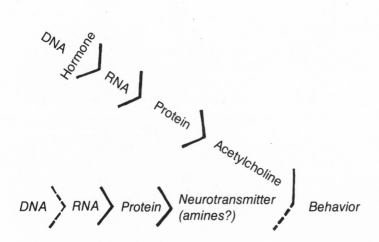

Figure 75.
Hypothetical relations between hormone induction of gene activity and territorial marking in the Mongolian gerbil.

marking to occur. Our one attempt to manipulate preoptic-area neurotransmitter levels in conjunction with hormone treatment supported the latter possibility. Atropine sulfate, an acetylcholine inhibitor, enhanced testosterone effects on marking, while pilocarpine nitrate, an acetylcholine stimulant, counteracted the hormone's effect.

Based on the results of the immunochemical and neurotransmitter studies, we have devised the model shown in Figure 75. The lower line depicts hypothetical processes which underlie territorial marking and are presumably necessary for the behavior to occur; these processes could be mediated by amines. The top line represents the synthesis of acetylcholine, which may inhibit marking by interfering with the basal processes shown in the lower line. The hormone may block the production of acetylcholine at the level of genetic transcription, thereby allowing marking to occur. The hormone could also disrupt genetic translation or other cytoplasmic processes.

The primary difference between this model and the one in Figure 74 is that here

the hormone inhibits, rather than induces, the formation of a neurotransmitter, which seems more consistent with the results of the studies presented in Chapter 5. According to this model, testosterone plus an inhibitor of acetylcholine could stimulate more marking than the hormone alone, because together they would produce greater inhibition of acetylcholine. Testosterone and a stimulant of acetylcholine action would presumably counteract each other, and this combined treatment, therefore, would not induce marking. It has been shown that drugs which alter catecholamine or serotonin levels inhibit the induction of marking by testosterone. Using the model in Figure 75, these drugs may inhibit processes shown in the lower line; actinomycin D, puromycin, and RNase may also act at this level. By this model it is even possible that the additional cellular component detected in the preoptic-area tissue of castrates could be the protein which regulates acetylcholine synthesis. The possibility that testosterone may act by suppressing acetylcholine formation is being studied further at this time.

Our results suggest not only that hormones activate behavior by stipulating neurological function, but also that they act directly on the species-specific nature of the organism. This action may involve the very basis of species-specificity, the transcription and translation of unique DNA-RNA sequences. Ultimately, we hope to specify the entire chain of events from gene induction to the occurrence of the territorial response under the modulating influence of olfactory cues. Because the gerbil's discrete scent marking behavior seems to be controlled by hormones acting on a specific brain site, the gerbil should be an ideal research animal for studying the neural physiology of hormone-dependent behaviors.

But what is the social significance of gerbil scent marks? Among insects, pheromones are chemical keys which activate genetically programed and often stereotyped behaviors. This rarely seems to be the case for mammals, though, and certainly is not true for gerbils. Our current information suggests that the chemical signal from the gerbil's scent gland is, for the most part, neutral until it is associated with environmental contingencies. The signal value, in other words, is contextual and can come to mean several things.

Concepts about pheromones should be revised to include such symbolic flexibility. The essential characteristic of a pheromone is not necessarily its molecular uniqueness or its ability to stimulate a particular response. Rather, the essential feature is its ability to be perceived by conspecifics and to be used by them to discriminate events. Almost any metabolic by-product might be used for information purposes as long as it provides consistent information about important environmental changes. Thus urinary or other metabolites that are natural by-products of physiological activity can be used as signals provided they accurately convey information about a conspecific's sex, social status, age, or other conditions. Similarly, the behavioral outcome can be quite variable, so long as it increases adaptability. Mammals, in particular, require a flexible system of communication and indeed would be at a disadvantage in many circumstances if they could not vary their behavior in regard to a signal. Similarly, they must be able to take note of any stimulus that can increase the total amount of information available. To be sure, there are genetic limitations to both the nature of the signal and the range of behavior that can be generated. Within these restrictions, however, there must be a great deal of flexibility.

Among gerbils the most common response to the scent gland secretion is attraction, but even this is variable and rapidly habituates. The pheromone retains important signal properties by being conditioned to environmental events. The signal, therefore, can probably be used to delimit territorial borders, to specify social class, to denote the presence or absence of relevant commodities, or to tag individuals or populations for recognition purposes. While we believe that the pheromone always works within a territorial context, we interpret this widely enough to include individual and population recognition, social class differences, reproductive capabilities, and aggressive functions.

Harderian secretion and spread during a facial groom also seem to relate to dominance status, aggressive functions, and possibly individual and population recognition. We know less about this pheromonal system than the ventral scent gland system, but they seem to correlate in most aspects and are used within the same territorial complex. The interesting suggestion arising out of the work with the Harderian gland is that pheromone emission, social interactions, and thermoregulatory processes are intimately associated. This relationship may extend to other pheromones, such as the ventral scent gland sebum.

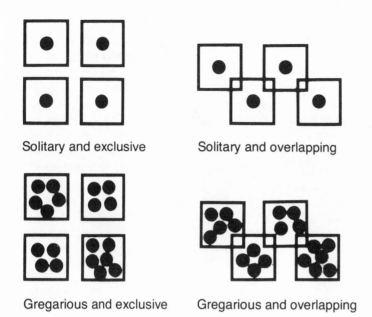

Solitary and exclusive   Solitary and overlapping

Gregarious and exclusive   Gregarious and overlapping

Figure 76.
Basic types of territorial systems (modified from Wynne-Edwards 1962).

Our notions about the type of territory gerbils acquire and defend are based on limited data. Four general types of territorial systems have been recognized by Wynne-Edwards (1962) and others, based on the number of animals occupying the territory and their neighbors. These four territory types are thus described as *solitary and exclusive, solitary and overlapping*, *gregarious and exclusive*, and *gregarious and overlapping*. They are diagrammed in Figure 76. Our best estimates, based on field observations and laboratory studies, indicate that gerbils form gregarious and exclusive territories. The family group seems to be the basic unit of territorial occupancy, although other females, adolescents, and subordinate males may occasionally coexist with the dominant family members.

The overall sequence of events which we feel characterizes territoriality in gerbils (see Table 23) includes three phases of territorial acquisition and defense: (*a*) a phase of social neutrality and attraction to conspecific sebum odors (*territorial advancement*), (*b*) a phase of intense fighting during which dominance is established (*territorial sweep*), and (*c*) a phase of social distinction, reinforced at least in part by olfactory signals (*territorial maintenance*). The social pressures that develop can cause submissive animals to migrate across ecological barriers, where possible. Such forced migration may in fact be a key to evolutionary diversification of populations and species. Migrants are animals that may establish new genetic populations independent of the parent population. They represent gene samples subject to new pressures of natural selection and as a result denote potentially new life styles. Territorial dominance ensures that the species will continually evolve in different directions.

Table 23.
Genesis and preservation of territorial behavior

Social Status	Behavioral States	Mechanisms
Ambiguous	*Territorial Advancement* Signal detection Attraction toward signal	*Directional modulators* Olfaction Locomotion
Developing	*Territorial sweep* Rapid and high-frequency marking High levels of fighting	*Oscillating modulators* Stimulus-bound Hormone-dependent
Clarified	*Territorial maintenance* A. Dominant animal: Periodic marking High mating success Winning of bouts Command of environment B. Submissive animal: Lowered marking Low mating success Losing of bouts Little command of environment Avoidance of olfactory stimuli of dominant class Migration	*Steady-state modulators* A. Dominant animal: Maintenance of hormone integrity B. Submissive animal: Decreased gonadotrophic secretion Increased stress reactions

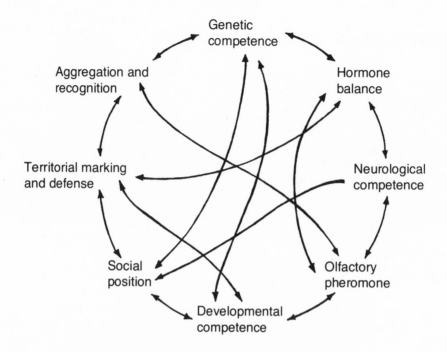

Figure 77.
Basic parameters of territoriality within the Central Integrative State.

In summary, the array of mechanisms that we feel regulate the territorial behavior of gerbils are depicted in Figure 77. They are both internal and external to the organism and are interrelated in complex ways. An animal must have the motor capabilities of scent marking its environment, an adequate hormonal status, and a sensitivity to the pheromonal qualities of the sebum. Whether or not the animal scent marks depends, however, on its social position and what the pheromone has come to mean under different conditions. In essence, a gerbil's territorial behavior depends upon the animal's genetic characteristics and all that this implies.

The next challenge facing us is to gather comparative data. We can then consider the possibility of building models of territoriality applicable to any species in which individuals compete for space and erect social barriers or migrate from the hub of the population.

Bibliography

Allen, G. M. 1940. Natural History of Central Asia. In *The mammals of China and Mongolia, Part II: Central Asiatic expeditions*, pp. 781–785. New York: American Museum of Natural History.

Anderson, K. M., and S. Liao. 1968. The selective retention of dihydrotestosterone by prostatic cell nuclei. *Nature* 219: 277–279.

Andrew, R. J. 1964. The displays of the primates. In *Evolutionary and genetic biology of primates*, edited by J. Buettner-Janusch, 1:227–309. New York: Academic Press.

Anisko, J. J., T. Christenson, and M. G. Buehler. 1973. Effects of androgen on fighting behavior in male and female Mongolian gerbils (*Meriones unguiculatus*). *Horm. & Behav.* 4:199–208.

Ardrey, R. 1966. *The territorial imperative*. New York: Atheneum Press.

Arluk, D. J. 1966. The effects of gonadal hormones on the ventral gland of the immature gerbil. *Amer. Zool.* 6:313.

—————. 1968. The hormonal regulation of the ventral sebaceous gland of the Mongolian gerbil. *Anatomy* 29-B.

Arrington, L. R., and C. B. Ammerman. 1969. Water requirements of gerbils. *Lab. Anim. Care* 10:503–505.

Arrington, L. R., T. C. Beaty, and K. C. Kelley. 1973. Growth, longevity and reproductive life of the Mongolian gerbil. *Lab. Anim. Sci.* 23(2):262–265.

Ashe, V. M., and G. McCain. 1972. Comparison of one-way and shuttle-avoidance of gerbils and rats. *J. Comp. & Physiol. Psych.* 80(2):293–296.

Bailey, V. 1936. *The mammals and life zones of Oregon*. North American Fauna 55. Washington, D.C.: U.S. Department of Agriculture.

Ball, J. 1939. Male and female mating behavior in prepubertally castrated male rats receiving estrogen. *J. Comp. & Physiol. Psych.* 28:273–283.

Balph, D. E., and A. W. Stokes. 1963. On the ethology of a population of Uinta ground squirrels. *Amer. Midl. Nat.* 69: 106–126.

Bannikov, A. G. 1954. The places inhabited and natural history of *Meriones unguiculatus*. In *Mammals of the Mongolian Peoples Republic*, pp. 410–415. USSR Academy of Sciences.

Baran, D. 1973. Responses of male Mongolian gerbils to male gerbil odors. *J. Comp. & Physiol. Psych.* 84:63–72.

Baran, D., and S. E. Glickman. 1970. Territorial marking in the Mongolian gerbil: A study of sensory control and function. *J. Comp. & Physiol. Psych.* 71:237–245.

Barash, D. P. 1973. Territorial and foraging behavior of pika (*Ochotona princeps*) in Montana. *Amer. Midl. Nat.* 89(1): 202–207.

Barfield, R. J. 1965. Induction of courtship and aggressive behavior by intracranial implants of androgen in capons. *Amer. Zool.* 5:203.

———. 1967. Activation of sexual and aggressive behavior by androgen implants in the brain of the male ring dove. *Amer. Zool.* 7:800.

Barfield, M. A., and E. A. Beeman. 1968. The oestrus cycle in the Mongolian gerbil, *Meriones unguiculatus*. *J. Reprod. & Fert.* 17:247–252.

Bauer, H. R. 1970. An experimental examination of the anti-predator response in the Mongolian gerbil (*Meriones unguiculatus*). M.A. thesis, University of Manitoba.

Beach, F. A. 1950. The snark was a boojum. *Amer. Psych.* 5:115–124.

———. 1970. Some effects of gonadal hormones on sexual behavior. In *The hypothalamus*, edited by L. Martini, M. Motta, and F. Fraschini, pp. 617–639. New York: Academic Press.

Berg, R. A., R. D. Shanin, and E. M. Hull. 1975. Early isolation in the gerbil (*Meriones unguiculatus*): Behavioral and physiological effects. *Physiol. Psych.* 3(1):35–38.

Beyer, C., P. McDonald, and N. Vidal. 1970. Failure of 5 α-dihydrotestosterone to elicit estrous behaviour in the ovariectomized rabbit. *Endocrinology* 86:939–941.

Billingham, R. S., and B. Seydian. 1963. Skin transplants and the hamster. *Sci. Amer.* 200:118–127.

Blass, E. M. 1965. The formation of learning sets by rodents. *Virginia J. Sci.* 1965: 403–404.

Blass, E. M., and A. R. Rollin. 1969. Formation of object discrimination learning sets by Mongolian gerbils (*Meriones unguiculatus*). *J. Comp. & Physiol. Psych.* 69:519–521.

Block, M. L., G. H. Vallier, and S. E. Glickman. 1974. Elicitation of water ingestion in the Mongolian gerbil (*Meriones unguiculatus*) by intracranial injections of angiotensin II and l-norepinephrine. *Pharmacol., Biochem. & Behav.* 2: 235–242.

Blum, S. L., and D. D. Thiessen. 1970. Effect of ventral gland excision on scent marking in the male Mongolian gerbil. *J. Comp. & Physiol. Psych.* 73:461–464.

———. 1971. The effect of different amounts of androgen on scent marking in the male Mongolian gerbil. *Horm. & Behav.* 2(4):279–285.

Boice, R., and J. G. Arledge. 1968. Water requirements of gerbils and kangaroo rats in the laboratory. *Psych. Reports* 23:1063–1069.

Boice, R., C. Boice, and A. E. Dunham. 1968. Role of docility in avoidance: Gerbils and kangaroo rats in a shuttlebox. *Psychon. Sci.* 10:381–382.

Boice, R., D. Hughes, and C. J. Cobb. 1969. Social dominance in gerbils and hamsters. *Psychon. Sci.* 16:127–128.

Boice, R., and A. Pickering. 1973. Shock-elicited behaviors in paired gerbils. *Psychon. Soc. Bull.* 2(4):221–223.

Bols, R. J., and R. Wong. 1973. Gerbils reared by rats: Effect on adult open field and ventral marking activity. *Behav. Biol.* 9(6):741–748.

Bourlière, F. 1970. *The natural history of mammals*. 3d ed. Translated by H. M. Parshley. New York: Alfred A. Knopf.

Britten, R. J., and E. H. Davidson. 1969. Gene regulation for higher cells: A theory. *Science* 165:349–357.

Brown, G. E. 1972. The effects of septal lesions in the Mongolian gerbil (*Meriones unguiculatus*). *Psychon. Sci.* 29(5):299–300.

Brown, L. E. 1966. Home range and movement of small mammals. *Symp. Zool. Soc. Lond.* 18:111–142.

Brownlee, R. G., R. M. Silverstein, D. Müller-Schwarze, and A. G. Singer. 1969. Isolation, identification and function of the chief component of the male tarsal scent in black-tailed deer. *Nature* 221:284–285.

Bruchovsky, N., and J. D. Wilson. 1968. The intranuclear binding of testosterone and 5 α-androstan-17β-01-3-one by rat prostate. *J. Biol. Chem.* 243:5953–5960.

Butler, C. G. 1966. Insect pheromones. *Rothamsted Experimental Station, Harpenden.* April 12.

Cabanac, M. 1974. Thermoregulatory behavior. In *Environmental physiology*, edited by D. Robertshaw. MTP International Review of Science, Physiology Series One VII. Baltimore: University Park Press.

———. 1975. Temperature regulation. In *Annual review of physiology*, vol. 37, edited by J. H. Comroe, Jr., R. R. Sonnenschein, and T. S. Edelman. Palo Alto: Annual Review.

Calhoun, J. B. 1963. *The ecology and sociology of the Norway rat.* U.S. Department of Health, Education, and Welfare, Public Health Service Publication, no. 1008. Washington, D.C.: Government Printing Office.

Campbell, N., D. Straney, and A. Neuringer. 1969. Operant conditioning in the Mongolian gerbil. *Psychon. Sci.* 16:255–256.

Carpenter, C. R. 1958. Territoriality: A review of concepts and problems. In *Behavior and evolution*, edited by A. Roe and G. G. Sympson, pp. 224–250. New Haven: Yale University Press.

Castell, R., and M. Maurus. 1967. The so-called urine washing of squirrel monkeys (*Saimiri scienurcus*) in dependence on environmental conditions and emotional factors. *Folia Primat.* 6: 170–176.

Chaworth-Musters, J. L., and J. E. Ellerman. 1947. A revision of the genus *Meriones*. *Proc. Zool. Soc. Lond.* 117:478–504.

Christenson, T., K. Wallen, B. A. Brown, and S. E. Glickman. 1973. Effects of castration, blindness, and anosmia on social reactivity in the male Mongolian gerbil (*Meriones unguiculatus*). *Physiol. & Behav.* 1:93–94.

Christian, J. J. 1970. Social subordination, population density and mammalian evolution. *Science* 168:84–90.

Clever, V. 1966. Gene activity patterns and cellular differentiation. *Amer. Zool.* 6:33–41.

Cole, J. M., and J. S. Topping. 1969. Depth and distance perception in the Mongolian gerbil. *Psychon. Sci.* 16:271–273.

Collins, A., G. Lindzey, and D. D. Thiessen. 1969. The regulation of cliff responses in the Mongolian gerbil (*Meriones unguiculatus*) by visual and tactual cues: I. *Psychon. Sci.* 16:227–229.

Crook, J. H. 1968. The nature and function of territorial aggression. In *Man and aggression*, edited by M. F. A. Montague, pp. 171–178. New York: Oxford University Press.

Cullen, J. W. 1972. Sodium intake in the Mongolian gerbil (*Meriones unguiculatus*) consequent to subcutaneous formalin injection. *Psychon. Sci.* 26: 279–282.

Cullen, J. W., and D. E. Scarborough. 1970. Behavioral and hormonal prophylaxis in the adrenalectomized gerbil (*Meriones unguiculatus*). *Horm. & Behav.* 1:203–210.

Dagg, A. I., and D. E. Windsor. 1971. Olfactory discrimination limits in gerbils. *Can. J. Zool.* 49:138–140.

Dambach, G. von. 1964. Die Morphologie der Glandula umbilicalis bei der mongolischen Rennmaus. *Aus der Abteilung für Experimentelle Pathologie Bd.* 4:116–121.

Darchen, R. 1964. Notes éthologiques sur le rat musque, *Ondatra Zibetica L.,* et en particulier sur la construction de la hutte d'hiver. *Mammalia* 28:137–168.

Davenport, R. K., Jr. 1967. The orangutan in Satah. *Folia Primat.* 5:247–263.

Davidson, J. M. 1966. Activation of the male's sexual behavior by intracranial implantation of androgen. *Endrocrinology* 79:783–794.

———. 1969. Effects of estrogen on the sexual behavior of male rats. *Endocrinology* 84:1365–1372.

Davidson, J. M., and S. Levine. 1972. Endocrine regulation of behavior. In *Annual review of physiology*, vol. 34, edited by J. H. Comroe, Jr., R. R. Sonnenschein, and A. C. B. Giese, pp. 375–408. Palo Alto: Annual Review.

Davis, S. F., W. P. Crutchfield, J. Shaver, and T. Sullivan. 1970. Interspecific odors as cues for runway behavior. *Psychon. Sci.* 20:166–167.

DeGhett, V. J. 1964. Behavioral development of the Mongolian gerbil (*Meriones unguiculatus*). *Ecol. Soc. Amer. Bull.* 50:90.

———. 1974. Developmental changes in the rate of ultrasonic vocalization in the Mongolian gerbil. *Develop. Psychobiol.* 7(3):267–273.

Dember, W. N., and R. Kleinman. 1973. Cues for spontaneous alternation by gerbils. *Anim. Learning & Behav.* 1(4):287–290.

Desjardins, C., J. A. Maruniak, and F. H. Bronson. 1973. Social rank in house mice: Differentiation revealed by ultraviolet visualization of urinary marking patterns. *Science* 182:939–941.

Dieterlen, F. 1959. Das Verhalten des syrischen Goldhamsters (*Mesocricetus auratus Waterhouse*): Untersuchungen zur Frage seiner Entwicklung und seiner angeborener Anteil durch geruschsisolierte Aufzuchten. *Z. Tierpsychol.* 16:47–103.

Dobroruka, L. J. 1960. Einige Beobachtungen an Ameisenigeln *Echidna aculeata Shaw* (1792). *Z. Tierpsychol.* 17:178–181.

Dobzhansky, T. 1937. *Genetics and the origin of species*. New York: Columbia University Press.

Doty, R. L., and R. Kart. 1972. A comparative and developmental analysis of the midventral sebaceous gland in 18 taxa of *Peromyscus* with an examination of gonadal steroid influences in *Peromyscus maniculatus bairdii*. *J. Mammal.* 53:83–99.

Doyle, G. A., A. Pelletier, and T. Bekker. 1967. Courtship, mating and parturition in the lesser bushbaby (*Galago senegalenses moholi*) under seminatural conditions. *Folia Primat.* 7:169–197.

Drickamer, L. C., and J. G. Vandenbergh. 1973. Predictors of social dominance in the adult female golden hamster. (*Mesocricetus auratus*). *Anim. Behav.* 21:564–570.

Drickamer, L. C., J. G. Vandenbergh, and D. R. Colby. 1973. Predictors of dominance in the male golden hamster (*Mesocricetus auratus*). *Anim. Behav.* 21:557–563.

Dücker, G. 1965. *Das Verhalten der Viverriden. Kukenthal: Handb. Zool. Berlin.* 8:1–48.

Dunstone, J. J., T. Cannon, J. T. Chickson, and W. K. Burns. 1972. Persistence and vigor of shock-induced aggression in gerbils (*Meriones unguiculatus*). *Psychon. Sci.* 28:272–274.

Dunstone, J. J., G. M. Krupski, and C. S. Weiss. 1971. Weight loss in gerbils (*Meriones unguiculatus*) continuously deprived of food, water, and both food and water. *Psych. Reports* 29:931–936.

Ebling, F. J. 1963. Hormonal control of sebaceous glands in experimental animals. In *Advances in biology of skin*, edited by W. Montagna, R. A. Ellis, and A. F. Silver, 4:200–219. New York: Macmillan.

Ebling, F. J., E. Elling, J. Skinner, and A. White. 1970. The response of the sebaceous glands of hypophysectomized-castrated male rats to adrenocorticotrophic hormone and to testosterone. *J. Endocrinol.* 48:73–81.

Eedy, J. W., and D. M. Ogilvie. 1970. The effect of age on the thermal preference of white mice (*mus musculus*) and gerbils (*Meriones unguiculatus*). *Can. J. Zool.* 48:1303–1306.

Eibl-Eibesfeldt, I. 1950. Über die Jugendentwicklung des Verhaltens eines mannlichen Dachses (*Meles meles L.*) unter besonderer Berücksichtigung des Spieles. *Z. Tierpsychol.* 7:327–355.

———. 1953a. Zur Ethologie des Hamsters (*Cricetus cricetus L.*). *Z. Tierpsychol.* 10:204–254.

———. 1953b. Vergleichende Studien an Ratten und Mausen. *Prakt. Desinfedtor* 45:166–168.

———. 1953c. Eine besondere Form des Duftmarkierens beim Riesengalago, *Galago crassicaudatus*. *Säugetierkdl. Mitt.* 1:171–173.

———. 1953d. Ethologische Unterschiede zwischen Hausratte und Wanderratte. *Zool. Anz. Suppl.* 16:169–180.

———. 1965. Das Duftmarkieren des Igeltanreks (*Echinops telfairi Martin*). *Z. Tierpsychol.* 22:810–812.

Eik-Nes, K. B. 1970. *The androgens of the testis*. New York: Marcel Dekker.

Eisenberg, J. F. 1962. Studies on the behavior of *Peromyscus maniculatus gambelii* and *Peromyscus californicus parasiticus*. *Behaviour* 19:177–207.

———. 1963. A comparative study in sandbathing behavior in heteromyid rodents. *Behaviour* 22:16–23.

———. 1967. A comparative study in rodent ethology with emphasis on evolution of social behavior: I. *Proc. U.S. Nat. Museum* 122:1–51.

Eisenberg, J. F., and D. G. Kleiman. 1972. Olfactory communication in mammals. In *Annual review of ecology and systematics*, vol. 3, edited by R. F. Johnston, P. W. Frank, and C. D. Michener, pp. 1–32. Palo Alto: Annual Review.

Eisenberg, J. F., and R. E. Kuehn. 1966. The behavior of *Ateles geoffroyi* and related species. *Smithsonian Misc. Coll.* 151:4683.

Epple, G. von. 1967. Comparative investigation on sexual and social behavior of Hapalidae. *Folia Primat.* 7:37–65.

———. 1970. Quantitative studies on scent marking in the marmoset (*Callithrix jacchus*). *Folia Primat.* 13:48–62.

———. 1974. Olfactory communication in South American primates. *Ann. N.Y. Acad. Sci.* 237(September):261–278.

Epple, G. von, and K. Lorenz. 1967. Distribution, morphology and function of the sternal glands in the platyrrhini. *Folia Primat.* 7:98–126.

Epsmark, Y. 1964a. Rutting behaviour in reindeer (*Rangifer tarandus L.*) *Anim. Behav.* 12:159–163.

———. 1964b. Studies in dominance, subordination relationships in a group of semi-domestic reindeer (*Rangifer tarandus L.*). *Anim. Behav.* 12:420–425.

Evans, C. S., and R. W. Goy. 1968. Social behavior and reproductive cycles in captive ringtailed lemurs (*Lemur catta*). *J. Zool. Lond.* 156:181–197.

Ewer, R. F. 1961. Further observations on suckling behaviour in kittens together with some general considerations of the interrelations of innate and acquired responses. *Behaviour* 18:247–260.

———. 1967. The behaviour of the African giant rat (*Cricetomys gambianus Waterhouse*). *Z. Tierpsychol.* 24:6–79.

———. 1968. *Ethology of mammals*. New York: Plenum Press.

———. 1971. The biology and behavior of a free-living population of black rats (*Rattus rattus*). *Anim. Behav. Monograph* 4:127–174.

Feldman, S., and N. Dafny. 1970. Effects of extrahypothalamic structures on sensory projections to the hypothalamus. In *The hypothalamus*, edited by L. Martini, M. Motta, and F. Fraschini, pp. 103–114. New York: Academic Press.

Ferguson, J. H., and G. E. Folk, Jr. 1970. The critical thermal minimum of small rodents in hypothermia. *Cryobiology* 7(1):44–46.

Finck, A., and H. Goehl. 1968. Vocal spectra and cochlear sensitivity in the Mongolian gerbil. *J. Audit. Res.* 8:63–69.

Finck, A., C. D. Schneck, and A. F. Hartman. 1972. Development of cochlear function in the neonate Mongolian gerbil (*Meriones unguiculatus*). *J. Comp. & Physiol. Psych.* 78(3):375–380.

Finck, A., and M. Sofouglu. 1966. Auditory sensitivity of the Mongolian gerbil. *J. Audit. Res.* 6:313–319.

Fisler, G. F. 1970. Communication systems and organizational systems in three species of rodents. *Bull. So. Calif. Acad. Sci.* 69:43–51.

Fradrich, H. 1967. The behavior of swine (*Seridae Tayassuidae*) and hippopotami (*Hippopotamidae*). *Handb. Zool.* 8:42.

Frank, F. 1956. Das Duftmarkieren der grossen Wühlmaus, *Arvicola terrestris (L.). Z. Säugetierk.* 21:172–175.

Frey, P., S. Eng, and W. Gavin. 1972. Instrumentation and techniques—conditioned suppression in the gerbil. *Behav. Res. Meth. & Instru.* 4(5):245–249.

Gaito, J. 1971. *DNA complex and adaptive behavior*. Englewood Cliffs, N.J.: Prentice-Hall.

Gallup, G. G., Jr., and M. S. Waite. 1970. Some preliminary observations on the behavior of Mongolian gerbils (*Meriones unguiculatus*) under seminatural conditions. *Psychon. Sci.* 20:25–26.

Galosy, R. A., and L. G. Lippman. 1970. Gerbil's pinnae movement as related to stimulus frequency and intensity. *Psychon. Sci.* 20:309–310.

Galosy, R. A., L. G. Lippman, and R. W. Thompson. 1970. Passive-avoidance learning in gerbils and rats. *J. Comp. & Physiol. Psych.* 73(1):269–273.

Galvani, P. F. 1971. The effects of partial reinforcement on the acquisition and extinction of avoidance behavior in gerbils. *Psychon. Sci.* 24:242–244.

Gee, E. P. 1953. Further observations on the great Indian one-horned rhinoceros. *J. Bombay Nat. Hist. Soc.* 51:765–772.

Gerritz, K. 1968. Sociability of male gerbils in the open-field. Paper presented at Eastern Psychological Association Annual Meeting.

Giegel, J. L., L. M. Stolfi, G. D. Weinstein, and P. Frost. 1971. Androgenic regulation of nucleic acid and protein synthesis in the hamster flank organ and other tissue. *Endocrinology* 89(3):904–909.

Gilbert, B. K. 1974. Scent marking and territoriality in pronghorn (*Antilocapra americana*) in Yellowstone National Park. *Mammalia* 37:25–33.

Ginsburg, H. J., and W. G. Braud. 1970. Changes in preference for cage environments following habituation and shock in the Mongolian gerbil. *Psychon. Sci.* 21:183–184.

———. 1971. A laboratory investigation of aggressive behavior in the Mongolian gerbil (*Meriones unguiculatus*). *Psychon. Sci.* 22:54–55.

Glasky, A. J., and L. N. Simon. 1966. Magnesium pemoline: Enhancement of brain RNA polymerase. *Science* 151: 702.

Gleason, K. K., and J. H. Reynierse. 1969. The behavioral significance of pheromones in vertebrates. *Psych. Bull.* 71:58–73.

Glenn, M. E., and J. Gray. 1964. Effect of various hormones on the growth and histology of the gerbil (*Meriones unguiculatus*) abdominal sebaceous gland pad. *Endocrinology* 76:1115–1123.

Glickman, S. E., L. Fried, and B. A. Morrison. 1967. Shredding of nest material in the Mongolian gerbil. *Percept. & Motor Skills* 24:473–474.

Glickman, S. E., and T. Higgins. 1968. Elicited behavior and reinforcement in the Mongolian gerbil. Paper presented at Symposium, Environment of Modification of Brain-Stimulation Effects, at Eastern Psychological Association Annual Meeting.

Glickman, S. E., T. Higgins, and R. L. Isaacson. 1970. Some effects of hippocampal lesions on the behavior of Mongolian gerbils. *Physiol. & Behav.* 5:931–938.

Goethe, F. 1938. Beobachtungen über das Absetzen von Witterungsmarken beim Baummarder. *Deut. Jager.* 13.

———. 1940. Beitrage zur Biologie des Iltis. *Z. Säugetierk.* 15:180–223.

Goldblatt, D. 1968. Seizure disorder in gerbils. *Neurology* 18:303–304.

Goldblatt, D., A. Konow, I. Shoulson, and T. MacMath. 1971. Effect of anticonvulsants on seizures in gerbils. *Neurology* 21:433–434.

Goodrich, B. S., and R. Mykytowycz. 1972. Individual and sex differences in the chemical composition of pheromone-like substances from the skin gland of the rabbit, *Oryctolagus cuniculus*. *J. Mammal.* 53:540–548.

Graf, W. 1956. Territorialism in deer. *J. Mammal.* 37:165.

Greenberg, G. 1973. Replication report: No spontaneous alternation in gerbils. *Psychon. Soc. Bull.* 1(2):141–143.

Griffo, W., and C. T. Lee. 1973. Progesterone antagonism of androgen-dependent marking in gerbils. *Horm. & Behav.* 4:351–358.

Grimsley, D. L. 1973. NaCl preference in the gerbil. *Physiol. Psych.* 1:93–94.

Gual, C., T. Morato, M. Huyano, M. Gut, and R. I. Dorfman. 1962. Biosynthesis of estrogens. *Endocrinology* 71:920–925.

Halpin, Z. T. 1974. Individual differences in the biological odors of the Mongolian gerbil (*Meriones unguiculatus*). *Behav. Biol.* 11:253–259.

Hamilton, T. H. 1968. Control by estrogen of genetic transcription and translation. *Science* 161:649–661.

Harriman, A. E. 1969a. Food and water requirements of Mongolian gerbils as determined through self-selection of diet. *Amer. Midl. Nat.* 82:149–156.

———. 1969b. A comparative study of food and water regulation by rats and Mongolian gerbils maintained on identical self-selection of diet schedules. *Amer. Midl. Nat.* 82:157–162.

Harris, G. G., and R. P. Michael. 1964. The activation of sexual behavior by hypothalamic implants of oestrogen. *J. Physiol.* 171:275–301.

Hart, B. L. 1973. Effects of castration on fighting, roaming, and urine spraying in adult male cats. *J. Amer. Vet. Med. Assoc.* 163(3):290–292.

———. 1974. Environmental and hormonal influences on urine marking behavior in the adult male dog. *Behav. Biol.* 11:167–176.

Hart, B. L., and C. M. Haugen. 1971. Scent marking and sexual behavior maintained in anosmic male dogs. *Commun. Behav. Biol.* 6:131–135.

Hart, J. S. 1971. Rodents. In *Comparative physiology of thermoregulation II: Mammals*, edited by G. C. Whittow, pp. 2–130. New York: Academic Press.

Harvey, E. B., and L. E. Rosenberg. 1960. An apocrine gland complex of the pika. *J. Mammal.* 41:213–219.

Hediger, H. 1949. Säugetier-Territorien und ihre Markierung. *Bijdragen tot de Dierkde* 28:172–184.

———. 1950. *Wild animals in captivity*. London: Butterworth and Co.

———. 1955. *Studies of the psychology of captive animals in zoos and circuses*. Translated by G. Sircom. New York: Criterion.

Hediger, H., and H. Kummer. 1956. Das Verhalten der Schnabeligel (*Tachyglossidae*). *Handb. Zool. Berl.* 8:1–8.

Higgins, T., S. E. Glickman, and R. Isaacson. 1967. The effects of hippocampal lesions on behavior patterns of the Mongolian gerbil, *Meriones unguiculatus*. *Psychon. Soc. Bull.* 1:26.

Hill, W. C. O. 1938. A curious habit common to Lorisoids and Platyrrhine monkeys. *Spolia Zeylonica* 21:65.

———. 1956a. Body odour in lorises. *Proc. Zool. Soc. Lond.* 127:580.

———. 1956b. Behavior and adaptation of the primates. *Proc. Royal Soc.* 66:94–110.

———. 1960. *Primates, IV: Cebidae; Part A*. Edinburgh: Edinburgh University Press.

Howard, H. E. 1920. *Territory in bird life*. New York: E. P. Dutton.

Howe, R. J. 1974. Marking behavior of the Bahamian hutia (*Geocapromys ingrahami*). *Anim. Behav.* 22(3):645–650.

Howell, A. B. 1926. Habits of the three subgenera studied. In *Anatomy of the wood rat*, edited by idem. Baltimore: Williams and Wilkins.

Hughes, R. E., and P. Nicholas. 1971. Effects of caging on the ascorbic acid content of the adrenal glands of the guinea pig and gerbil. *Life Sci.* 10(pt. 2):53–55.

Hull, E. M., K. L. Hamilton, D. B. Engwall, and L. Rosselli. 1974. Effects of olfactory bulbectomy and peripheral deafferentiation on reactions to crowding in gerbils (*Meriones unguiculatus*). *J. Comp. & Physiol. Psych.* 86:247–254.

Hull, E. M., C. J. Langan, and L. Rosselli. 1973. Population density and social, territorial, and physiological measures in gerbils. *J. Comp. & Physiol. Psych.* 84:414–422.

Illiger, J. K. 1811. *Meriones. Prodromus Syst. Mamm. et Avian.*, p. 82. SEE Schwentker, V., 1968.

Ilse, D. 1955. Olfactory marking of territory in two young male loris, *Loris Tardigradus lydekkerianus*, kept in captivity in Poona. *Brit. J. Anim. Behav.* 3:118–120.

Ireland, L. C., and R. L. Isaacson. 1968. Reactivity in the hippocampectomized gerbil. *Psychon. Sci.* 12:163–164.

Jakinovich, W., and B. Oakley. 1975. Comparative gustatory responses in four species of gerbillinae rodents. *J. Comp. Physiol.* 99:89–101.

Jarbe, T. U. C., J. O. Johansson, and B. G. Henriksson. 1975. Δ^9-tetrahydrocannabinol and pentobarbital as discriminative cues in the Mongolian gerbil (*Meriones unguiculatus*). *Pharmacol., Biochem. & Behav.* 3:403–410.

Jettmar, H. M. 1930. Biologische Beobachtungen über einige Nager im süd-
mandschurish-mongolischen Grenzge-
biet. *Z. Säugetierk.* 5:344–361.

Johnston, R. E. 1969. Scent marking in
male golden hamsters. Paper presented at Eastern Psychological Association Annual Meeting.

———. 1970. Scent marking in female
hamsters. Paper presented at Eastern
Psychological Association Annual
Meeting.

———. 1975*a*. Scent marking by male
golden hamsters (*Mesocricetus auratus*), I: Effects of odors and social encounters. *Z. Tierpsychol.* 37:75–98.

———. 1975*b*. Scent marking by male
golden hamsters (*Mesocricetus auratus*), II: The role of the flank gland scent
in the causation of marking. *Z. Tierpsychol.* 37:138–144.

———. 1975*c*. Scent marking by male
golden hamsters (*Mesocricetus auratus*), III: Behavior in a seminatural environment. *Z. Tierpsychol.* 37:213–221.

Jolly, A. 1966. *Lemur behavior: A Madagascar field study*. Chicago: University
of Chicago Press.

Jones, R. B., and N. W. Nowell. 1973. Effects of preputial and coagulating
gland secretions upon aggressive behaviour in male mice: A confirmation.
J. Endocrinol. 59:203–204.

Kalabukhov, N. I. 1964. Effect of vitamins:
E (tocophero 1) and C (ascorbic acid)
on hibernating rodents. *Byulleten
Moskovskogo Obshchestva Isptatilei
Prifody, Otdel Biologicheskii* 69:15.
Translation in *Federation Proc.* 24:851–
857 (1964).

Kaplan, H., and S. O. Hyland. 1972. Behavioral development in the Mongolian
gerbil (*Meriones unguiculatus*). *Anim.
Behav.* 20(1):147–154.

Kaplan, H., and C. Miezejeski. 1972. Development of seizures in the Mongolian
gerbil (*Meriones unguiculatus*). *J.
Comp. & Physiol. Psych.* 81(2):267–
273.

Kaufmann, J. H. 1965. A three-year study of
mating behavior in a free-ranging band
of rhesus monkeys. *Ecology* 30:146–
155.

Kaufmann, J. H., and A. Kaufmann. 1965.
Observations of the behavior of tayras
and grisons. *Z. Säugetierk.* 30:146–
155.

King, J. E., R. R. Goodman, and W. W.
Rees. 1968. Two- and four-choice
object discrimination by gerbils. *J.
Genet. Psych.* 112:117–125.

Kivett, V. K. 1975. Variations in integumen-
tary gland activity and scent marking
in Columbian ground squirrels (*Spermophilus c. columbianus*). Thesis,
University of Alberta.

Kleiman, D. G. 1974. Scent marking in the
binturong, *Arctictis binturong. J. Mammal.* 55(1):224–227.

Koenig, L. 1960. Das Aktionssystem des
Siebenschläfers (*Glis glis L.*). *Z. Tierpsychol.* 17:427–505.

Komisaruk, B. R. 1967. Effects of local brain
implants of progesterone on reproductive behavior in ring doves. *J. Comp.
& Physiol. Psych.* 64:219–224.

Komisaruk, B. R., and C. Beyer. 1972.
Responses of diencephalic neurons to
olfactory bulb stimulation, odor, and
arousal. *Brain Res.* 36:153–170.

Kramer, A. 1970. Social organization and
social behavior in chamois population
(*Rupicapra rupicapra L.*) of the alps.
Z. Tierpsychol. 26:889–964.

Kramis, R. C., and A. Routtenberg. 1969.
Rewarding brain stimulation, hippocampal activity, and foot-stomping in
the gerbil. *Physiol. & Behav.* 4:7–11.

Kuehn, R. E., and I. Zucker. 1966. Mating behavior of *Meriones unguiculatus*. *Amer. Zool.* 6:535.

————. 1968. Reproductive behavior of the Mongolian gerbil (*Meriones unguiculatus*). *J. Comp. & Physiol. Psych.* 66:747–752.

Kuhme, W. 1961. Beobachtungen am Afrikanischen Elefanten (*Loxodonta african Blumenbach*, 1797) in Gefangenschaft. *Z. Tierpsychol.* 18:285–296.

Kunkel, P., and I. Kunkel. 1964. Beitrage zur Ethologie des Hausmeerschweinchens, *Cavia aperea f. porcellus (L.)*. *Z. Tierpsychol.* 21:603–641.

Kutscher, C. L. 1969. Species differences in the interaction of feeding and drinking. *Ann. N.Y. Acad. Sci.* 157(2):539–552.

Kutscher, C. L., R. D. Stillman, and I. Weiss. 1968. Food deprivation polydipsia in gerbils (*Meriones unguiculatus*). *Physiol. & Behav.* 3:667–671.

Lack, D. 1966. *Population studies of birds*. London: Oxford University Press.

Laing, D. G. 1975. A comparative study of the olfactory sensitivity of humans and rats. *Chemical Senses and Flavor* 1:257–269.

Lee, C. T., and D. Estep. 1971. The developmental aspect of marking and nesting behavior in Mongolian gerbils (*Meriones unguiculatus*). *Psychon. Sci.* 22:312–313.

Lerwill, C. J. 1974. Activity rhythms of golden hamsters (*Mesocricetus auratus*) and Mongolian gerbils (*Meriones unguiculatus*) by direct observation. *J. Zool.* 174:520–523.

Leyhausen, P., and R. Wolff. 1959. Das Revier einer Hauskatze. *Z. Tierpsychol.* 16:666–670.

Lindzey, G., D. D. Thiessen, and A. Tucker. 1968. Development and hormonal control of territorial marking in the male Mongolian gerbil (*Meriones unguiculatus*). *Develop. Psychobiol.* 1:97–99.

Linsdale, J. M. 1946. *The California ground squirrel: A record of observations made on the Hastings Natural History Reservation*. Berkeley and Los Angeles: University of California Press.

Linsdale, J. M., and L. P. Levis. 1951. *The dusky-footed wood rat: A record of observations made on the Hastings Natural History Reservation*. Berkeley and Los Angeles: University of California Press.

Lipkow, J. 1954. Über des Seitenorgan des Goldhamsters (*Mesocricetus a. auratus Waterhouse*). *Z. Morph. u. Okol. d. Tiere*. 42:333–372.

Lippman, L. G., and R. A. Galosy. 1969. Pinnae movement as related to auditory intensity in gerbils. *Psychon. Sci.* 16(5):276.

Lippman, L. G., R. A. Galosy, and R. W. Thompson. 1970. Passive-avoidance learning in gerbils and rats. *J. Comp. & Physiol. Psych.* 73:269–273.

Lisk, R. D. 1962. Diencephalic placement of estradiol and sexual receptivity in the female rat. *Amer. J. Physiol.* 203:493–496.

Lockard, R. B. 1968. The albino rat: A defensible choice or a bad habit? *Amer. Psych.* 23:734–742.

Lorenz, K. 1950. The comparative method in studying innate behaviour patterns. *Symp. Soc. Exp. Biol.* 4:221–268.

————. 1966. *On aggression*. New York: Harcourt, Brace and World.

Loskota, W. J., P. Lomax, and S. T. Rich. 1972. The gerbil as a model for the study of epilepsy: Seizure habituation and seizure patterns. *Proc. West Pharmacol. Soc.* 15:189–194.

————. 1974. The gerbil as a model for the study of the epilepsies. *Epilepsia* 15:109–119.

Loskota, W. J., P. Lomax, and M. A. Verity. 1974. *A stereotaxic atlas of the Mongolian gerbil brain (*Meriones unguiculatus*)*. Ann Arbor: Ann Arbor Science Publishers.

Ludvigson, H. W., and D. Sytama. 1964. The sweet smell of success: Apparent double alternation in the rat. *Pschon. Sci.* 9:283–284.

Luttge, W. G., and R. E. Whalen. 1970. Dihydrotestosterone, androstenedione, testosterone: Comparative effectiveness in masculinizing and defeminizing reproductive systems in male and female rats. *Horm. & Behav.* 1:265–281.

McClintock, M. K. 1971. Menstrual synchrony and suppression. *Nature* 229:244–245.

McDonald, R., C. Beyer, F. Newton, B. Brien, R. Baker, H. S. Tan, C. Sampson, P. Kitching, R. Greenhill, and D. Pritchard. 1970. Failure of 5 α-dihydrotestosterone to initiate sexual behavior in the castrated male rat. *Nature* 227:964–965.

McHose, J. H. 1967. Patterned running as a function of the sequence of trial administration. *Psychon. Sci.* 9:281–282.

McHose, J. H., and H. W. Ludvigson. 1966. Differential conditioning with nondifferential reinforcement. *Psychon. Sci.* 6:485–486.

McManus, J. 1971. Early postnatal growth and the development of temperature regulation in the Mongolian gerbil (*Meriones unguiculatus*). *J. Mammal.* 52:782–792.

Marsden, H. M., and N. R. Holler. 1964. Social behavior in confined populations of the cottontail and the swamp rabbit. *Wildl. Mongr. Chestertown* 13:6–39.

Marston, J. H., and M. C. Chang. 1965. The breeding, management and reproductive physiology of the Mongolian gerbil (*Meriones unguiculatus*). *Lab. Anim. Care* 15:34–48.

Martin, R. C., E. Ragland, and K. B. Melvin. 1970. Self-punitive locomotor behavior in the Mongolian gerbil. *Psychon. Sci.* 20:183–184.

Mason, W. A. 1966. Social organization of the South American monkey, *Callicebus moloch*: A preliminary report. *Tulane Studies Zool.* 13:23–28.

Mayr, E. 1970. *Populations, species and evolution*. Cambridge: Harvard University Press.

Meckley, P. E., and O. J. Ginther. 1972. Effects of litter and male on corpora lutea of the postpartum Mongolian gerbil. *J. Anim. Sci.* 34:297–301.

Michael, R. P., E. B. Keverne, and R. W. Bonsall. 1971. Pheromones: Isolation of male sex attractants from female primate. *Science* 172:964–966.

Michael, R. P., and D. Zumpe. 1970. Sexual initiating behavior by female rhesus monkeys (*Macaca mulatta*) under laboratory conditions. *Behaviour* 36:168–186.

Milne-Edwards, A. 1867. Observations sur quelques mammifères du Nord de la Chine. *Ann. Sci. Nat. (Zool.)* 7:375–377.

Milner, J. 1972. Anticholingeric blockade of water intake in the Mongolian gerbil. *Psychon. Sci.* 26:135–136.

Mitchell, O. G. 1965. Effect of castration and transplantation on ventral gland of the gerbil. *Proc. Soc. Exp. Biol. Med.* 119:953–955.

———. 1967. The supposed role of the gerbil ventral gland in reproduction. *J. Mammal.* 48:142.

Moncrieff, R. W. 1951. *The chemical senses*. London: Leonard Hill.

Montagna, W. 1962. *The structure and function of skin*. 2d ed. New York: Academic Press.

Morris, D. 1962. The behaviour of the green acouchi (*Myoprocta pratti*) with special reference to scatter hoarding. *Proc. Zool. Soc. Lond.* 139:701–732.

———. 1964. *The mammals: A guide to the living species*. New York: Harper and Row.

Moy, F. F. 1970. Histology of the subauricular and rump glands of the pronghorn (*Antilocapra americana Ord*). *Amer. J. Anat.* 129:65–88.

Moynihan, M. 1966. Communication in *Callicebus. J. Zool. Lond.* 150:77–127.

———. 1967. Comparative aspects of communication in New World primates. In *Primate ethology*, edited by D. Morris, pp. 236–266. Chicago: Aldine.

Müller-Schwarze, D. 1967. Social odors in young male deer. *Amer. Zool.* 7:430.

———. 1969a. Pheromone function of deer urine. *Amer. Zool.* 9:21.

———. 1969b. Complexity and relative specificity in a mammalian pheromone. *Nature* 223:525–526.

———. 1971. Pheromones in blacktail deer (*Odocoileus hemionus columbianus*). *Anim. Behav.* 19:141–152.

Müller-Schwarze, D., and C. Müller-Schwarze. 1969. A herd of blacktail deer. *Pacific Disc.* 22:22–26.

Müller-Using, D. 1956. Zum Verhalten des Murmeltieres (*Marmota marmota* [*L.*]). *Z. Tierpsychol.* 13:135–142.

Murphy, M. R. 1970a. Olfactory bulb removal reduces social territorial behavior in the male golden hamster. Paper presented at Eastern Psychological Association Annual Meeting.

———. 1970b. Territorial behavior of the caged golden hamster. *Proceedings of the 78th Annual Convention, American Psychological Association*. Washington, D.C.: APA.

———. 1971. Olfaction and territorial behavior of the caged golden hamster. Paper presented at Eastern Psychological Association Annual Meeting.

Myers, K., and W. E. Poole. 1961. A study of the biology of the wild rabbit, *Oryctolagus cuniculus (L.)* in confined populations, II: The effects of season and population increase on behaviour. *CSIRO Wildl. Res.* 6:1–41.

Mykytowycz, R. 1965. Further observations on the territorial function and histology of the submandibular cutaneous (chin) glands in the rabbit, *Oryctolagus cuniculus (L.). Anim. Behav.* 8:400–412.

———. 1966a. Observations on odoriferous and other glands in the Australian wild rabbit, *Oryctolagus cuniculus (L.)*, and the hare, *Lepus europaeus P.*, I: The anal gland. *CSIRO Wildl. Res.* 11:11–29.

———. 1966b. Observations on odoriferous and other glands in the Australian wild rabbit, *Oryctolagus cuniculus (L.)*, and the hare, *Lepus europaeus P.*, II: The inguinal glands. *CSIRO Wildl. Res.* 11:29–64.

———. 1966c. Observations on odoriferous and other glands in the Australian wild rabbit, *Oryctolagus cuniculus (L.)*, and the hare, *Lepus europaeus P.*, III: Harder's lachrymal and submandibular glands. *CSIRO Wildl. Res.* 11:65–90.

———. 1968. Territorial marking by rabbits. *Sci. Amer.* 218:116–126.

———. 1970. The role of skin glands: Mammalian communication. In *Communication by chemical signals*, edited by J. W. Johnston, Jr., D. G. Moulton, and A. Turk, pp. 327–360. New York: Appleton-Century-Crofts.

———. 1974. Odor in the spacing behavior of mammals. In *Pheromones*, edited by M. C. Birch, pp. 327–343. Amsterdam: North Holland Pub. Co.

Mykytowycz, R., and S. Gambale. 1969. The distribution of dung-hills and the behavior of free-living wild rabbits, *Oryctolagus cuniculus (L.)*, on them. *Forma et Functio* 1:333.

Nauman, D. J. 1963. The Mongolian gerbil as an experimental animal in behavioral research. Paper presented at North and South Dakota Bi-State Psychological Convention, Aberdeen, South Dakota.

———. 1968. Open field behavior of the Mongolian gerbil. *Psychon. Sci.* 10: 163–164.

Neumann, F. 1971–1972. Use of cyproterone acetate in animal and clinical trials. *Horm. & Antag. Gynec. Invest.* 2:150–179.

Nichol, A. A. 1938. *Experimental feeding of deer*. University of Arizona College Agricultural Experimental Station, Technical Bulletin 75. Tucson.

Nikkari, T., and M. Valavaara. 1969. The production of sebum in young rats: Effects of age, sex, hypophysectomy and treatment with somatotrophic hormone and sex hormones. *J. Endocrinol.* 43:113–118.

Nolte, A. 1958. Beobachtungen über das Instinktverhalten von, Kapuzinerafferen (*Cebus apella*) in der Gefangenschaft. *Behaviour* 12:183–207.

Norris, M. L., and C. E. Adams. 1971. Delayed implantation in the Mongolian gerbil (*Meriones unguiculatus*). *J. Reprod. & Fert.* 27:486–487.

———. 1972. Aggressive behavior and reproduction in the Mongolian gerbil (*Meriones unguiculatus*) relative to age and sexual experience at pairing. *J. Reprod. & Fert.* 31:447–450.

Nyby, J., J. K. Belknap, and D. D. Thiessen. 1974. The effects of d- and l-amphetamine upon hoarding behavior and feeding in the Mongolian gerbil (*Meriones unguiculatus*). *Physiol. Psych.* 2(4):497–499.

Nyby, J., and D. D. Thiessen. 1971. Singular and interactive effects of testosterone and estrogen on territorial marking in castrated male Mongolian gerbils (*Meriones unguiculatus*). *Horm. & Behav.* 2:279–285.

Nyby, J., D. D. Thiessen, and P. Wallace. 1970. Social inhibition of territorial marking in the Mongolian gerbil (*Meriones unguiculatus*). *Psychon. Sci.* 21: 310–312.

Nyby, J., P. Wallace, K. Owen, and D. D. Thiessen. 1973. An influence of hormones upon hoarding behavior in the Mongolian gerbil (*Meriones unguiculatus*). *Horm. & Behav.* 4:283–288.

Oldham, J., and H. Morlock. 1970. The effects of openfield size on activity in the Mongolian gerbil. *Psychon. Sci.* 20:290.

O'Malley, B. W., and A. R. Means. 1974. Female steroid hormones and target cell nuclei. *Science* 183:610–620.

Owen, K. 1972. The hormonal control of scent marking, sex behavior and ovulation in the Mongolian gerbil (*Meriones unguiculatus*). Ph.D. dissertation, University of Texas at Austin.

Owen, K., and D. D. Thiessen. 1973. Regulation of scent marking in the female Mongolian gerbil (*Meriones unguiculatus*). *Physiol. & Behav.* 11:441–445.

Owen, K., P. Wallace, and D. D. Thiessen. 1974. Effects of intracerebral implants of steroid hormones on scent marking in the ovariectomized female gerbil (*Meriones unguiculatus*). *Physiol. & Behav.* 12:755–760.

Palka, Y. S., and C. H. Sawyer. 1966. The effects of hypothalamic implants of ovarian steroids on estrous behavior in rabbits. *J. Physiol.* 180:251–269.

Pav, D., and S. I. Magalini. 1966. Studies on the *Meriones unguiculatus* (Mongolian gerbil). *Metabolismo* 2:137–149.

Petter, J. J. 1962a. Ecological and behavioural studies of Madagascar lemurs in the field. *Ann. N.Y. Acad. Sci.* 102:267–281.

———. 1962b. Recherches sur l'écologie et l'éthologie des lemuriens malgaches. *Mem. Mus. Nat. Hist., Nat. Ser. A* 27:146.

———. 1965. The lemurs of Madagascar. In *Primate behavior*, edited by I. Devore, pp. 292–319. New York: Holt, Rinehart and Winston.

Petter, J. J., and J. Petter-Rousseaux. 1956. A propos du lemurien malgache *cheirogales trichotis. Mammalia* 20: 46–48.

Petter-Rousseaux, J. 1964. Reproductive physiology and behavior of the lemuroides. In *Evolutionary and genetic biology of primates*, edited by J. Buettner-Janusch, 1:91–132. New York: London.

Pfeiffer, W. 1962. The fright reaction of fish. *Biol. Rev.* 37:495–511.

Phoenix, C. H., R. W. Goy, and W. C. Young. 1967. Sexual behavior: General aspects. In *Neuroendocrinology*, edited by L. Martini and W. F. Ganong, pp. 163–196. New York: Academic Press.

Pilters, H. 1956. Das Verhalten der Tylopoden. *Handb. Zool. Berl.* 8:1–24.

Pocock, R. I. 1910. On the specialised cutaneous glands of ruminants. *Proc. Zool. Soc. Lond.* 1910 (June):840–986.

Powell, R. W. 1971. Acquisition of free-operant (Sidman) avoidance in Mongolian gerbils (*Meriones unguiculatus*) and albino rats. *Psychon. Sci.* 22:279–281.

———. 1972. Analysis of warm-up effects during avoidance in wild and domesticated rodents. *J. Comp. & Physiol. Psych.* 78:311–316.

Powell, R. W., and S. Peck. 1969. Running-wheel activity and avoidance in the Mongolian gerbil. *J. Exp. Anal. Behav.* 12:779–787.

Prior, R. 1968. *The roe deer of Cranborne Chase.* London: Oxford University Press.

Quay, W. B. 1953. Seasonal and sexual differences in the skin gland of the kangaroo rat (*Dipodomys*). *J. Mammal.* 34:1–14.

Raisman, G. 1966. Neural connections of the hypothalamus. *Brit. Med. Bull.* 22(3):197–201.

———. 1970. Some aspects of the neural connections of the hypothalamus. In *The hypothalamus*, edited by L. Martini, M. Motta, and F. Fraschini. New York: Academic Press.

Ralls, K. 1971. Scent marking in captive Maxwell's duikers. In *The papers of An International Symposium on the Behaviour of Ungulates and Its Relation to Management.* Calgary: University of Calgary.

Reynierse, J. H. 1971. Agonistic behavior in Mongolian gerbils. *Z. Tierpsychol.* 29:175–179.

Reynierse, J. H., M. J. Scavio, Jr., and J. D. Ulness. 1969. Gerbil runway performance under hunger motivation. *Psychon. Sci.* 16:36–37.

Rice, M. 1975. Relationships among sex behavior, scent marking, aggression and plasma testosterone in the male Mongolian gerbil (*Meriones unguiculatus*). Ph.D. dissertation, University of Texas at Austin.

Rich, S. T. 1968. The Mongolian gerbil (*Meriones unguiculatus*) in research. *Lab. Anim. Care* 18:235–243.

Rieder, C. A., and J. H. Reynierse. 1971. Effects of maintenance condition on aggression and marking behavior of the Mongolian gerbil (*Meriones unguiculatus*). *J. Comp. & Physiol. Psych.* 75:471–475.

Robinson, D. G., Jr. 1968. Animals suited to epileptic research. *Sci. News* 93: 16–18.

———. 1969. Friendly desert jumpers. *Highlights for Children* (January).

Rosenzweig, M. R., and E. L. Bennett. 1969. Effects of differential environments on brain weights and enzyme activities in gerbils, rats and mice. *Develop. Psychobiol.* 2:87–95.

Rothenberg, D. 1974. Territorial behavior characteristics in gerbils. *Psych. Reports* 34:810.

Roth-Kolar, H. 1957. Beiträge zu einem Aktionssystem des Aguti (*Dasyprocta aguti aguti L.*). *Z. Tierpsychol.* 14: 362–375.

Routtenberg, A., and R. C. Kramis. 1967. "Foot-stomping" in the gerbil: Rewarding brain stimulation, sexual behavior and foot shock. *Nature* 214:173–174.

Ryan, K. J. 1960. Estrogen formation by the human placenta: Studies on the mechanism of steroid aromatization by mammalian tissue. *Acta Endocrinol.* 35:697–698 (Suppl. 51).

Sayler, A. 1970. The effect of anti-androgens on aggressive behavior in the gerbil. *Physiol. & Behav.* 5:667–671.

Schaffer, J. 1940. *Die Hautdrüsenorgane der Säugetiere*. Berlin: Wien.

Schnurr, R. 1971. Spontaneous alternation in normal and brain-damaged gerbils. *Psychon. Sci.* 25(3):181–182.

Schultz-Westrum, T. 1965. Innerartliche Verständigung durch Dufte beim Gleitbeutler *Petaurus breviceps papuanus Thomas* (Marsupilia Phalangeridae). *Z. Vergleich. Physiol.* 50:151–220.

———. 1969. Social communication by chemical signals. In *Olfaction and taste*, edited by C. Pfaffmann. New York: Rockefeller University Press.

Schwentker, V. 1965. The gerbil: A new laboratory animal. *Ill. Vet.* 6:5–9.

———. 1968. *The gerbil: An annotated bibliography*. Brant Lake, N.Y.: Tumblebrook Farm.

Scott, J. P., and J. L. Fuller. 1965. *Genetics and the social behavior of the dog*. Chicago: University of Chicago Press.

Scott, J. W., and C. M. Leonard. 1971. The olfactory connections of the lateral hypothalamus in the rat, mouse and hamster. *J. Comp. Neurol.* 141:331–344.

Sewell, G. D. 1970. Ultrasonic signals from rodents. *Ultrasonics* 8:26–30.

Sharp, P. L. 1973. Behavior of the pika (*Ochotona princeps*) in the Kananaskis Region of Alberta. Thesis, University of Alberta.

Sikes, S. 1964. The ratel or honey badger. *Afr. Wildl.* 16:275–281.

Sleggs, G. 1926. The adult anatomy and histology of the anal glands of the Richardson ground-squirrel *Citellus richardsonii* Sabine. *Anat. Rec.* 32:1–43.

Smith, R. L., and S. Zwislocki. 1971. Responses of some neurons of the cochlear nucleus to tone increments. *J. Acoustical Soc. Amer.* 150:1520–1525.

Sobell, H. M., S. C. Jain, T. D. Sakore, and C. E. Nordman. 1971. Stereochemistry of actinomycin-DNA binding. *Nature New Biol.* 231:200–205.

Sokolov, W., and L. Skurat. 1966. A specific midventral gland in gerbils. *Nature* 211:544–545.

Sprankel, H. 1961. On the behavior and breeding of *Tupaia glis* (Diard, 1820) in captivity. *Z. für Wiss. Zool.* 165: 187–220.

Steiner, A. L. 1973. Self- and allo-grooming behavior in some ground squirrels (Sciuridae), a descriptive study. *Can. J. Zool.* 51(2):151–161.

———. 1974. Body rubbing, marking and other scent-related behavior in some ground squirrels (Sciuridae), a descriptive study. *Can. J. Zool.* 52:889–906.

Strauss, J. S., and F. J. Ebling. 1970. Control and function of skin glands in mammals. In *Hormones and the environment*, edited by G. K. Benson and J. G. Phillips, pp. 341–368. Cambridge: The University Press.

Strauss, J. S., and P. E. Pochi. 1963. The human sebaceous gland: Its regulation by steroidal hormones and its use as an end organ for assaying androgenicity *in vivo*. *Recent Prog. Horm. Res.* 19:385.

Stutz, A. 1971. Effects of weak magnetic fields on gerbils' spontaneous activity. *Ann. N.Y. Acad. Sci.* 188:312–323.

Tanimoto, K. 1943. Ecological studies on plague-carrying animals in Manchuria. *Dobutsugaku Zasshi* [Zool. Mag.] 55:111–127.

Tata, J. R. 1966. Hormones and synthesis and utilization of ribonucleic acids. *Prog. Nucleic Acid Res.* 5:191–250.

Tavolga, W. N. 1956. Visual, chemical and sound stimuli as cues in the sex discriminatory behavior of the Gobiid fish, *Bathygobius saparator. Zoologica* 41: 49–64.

Tembrock, G. 1968. Land mammals. In *Animal communication: Techniques of study and results of research*, edited by T. A. Sebeok, pp. 338–404. Bloomington: Indiana University Press.

Thiessen, D. D. 1964. Population density and behavior: A review of theoretical and physiological contributions. *Texas Rep. Biol. Med.* 22:266–314.

———. 1968. The roots of territorial marking in the Mongolian gerbil: A problem of species-common topography. *Behav. Res. Meth. & Instru.* 1:70–76.

———. 1973. Footholds for survival. *Amer. Sci.* 71:346–351.

Thiessen, D. D., S. Blum, and G. Lindzey. 1969. A scent marking response associated with the ventral sebaceous gland of the Mongolian gerbil (*Meriones unguiculatus*). *Anim. Behav.* 18:26–30.

Thiessen, D. D., A. Clancy, and M. Goodwin. 1976. Harderian pheromone in the Mongolian gerbil (*Meriones unguiculatus*). *J. Chem. Ecol.*

Thiessen, D. D., and M. Dawber. 1972. Territorial exclusion and reproductive isolation. *Psychon. Sci.* 28:158–160.

Thiessen, D. D., H. C. Friend, and G. Lindzey. 1968. Androgen control of territorial marking in the Mongolian gerbil. *Science* 160:26–30.

Thiessen, D. D., and S. Goar. 1970. Stereotaxic atlas of the hypothalamus of the Mongolian gerbil (*Meriones unguiculatus*). *J. Comp. Neurol.* 140:123–128.

Thiessen, D. D., M. Goodwin, and A. Clancy. 1976. Harderian gland: A behavioral puzzle. Unpublished.

Thiessen, D. D., and G. Lindzey. 1970. Territorial marking in the female Mongolian gerbil: Short-term reactions to hormones. *Horm. & Behav.* 1:157–160.

Thiessen, D. D., G. Lindzey, S. L. Blum, A. Tucker, and H. C. Friend. 1968. Visual behavior of the Mongolian gerbil (*Meriones unguiculatus*). *Psychon. Sci.* 11:23–24.

Thiessen, D. D., G. Lindzey, S. L. Blum, and P. Wallace. 1970. Social interactions and scent marking in the Mongolian gerbil (*Meriones unguiculatus*). *Anim. Behav.* 19:505–513.

Thiessen, D. D., G. Lindzey, and A. Collins. 1969. Early experience and visual cliff behavior in the Mongolian gerbil (*Meriones unguiculatus*): II. *Psychon. Sci.* 16:240–241.

Thiessen, D. D., G. Lindzey, and H. C. Friend. 1968. Spontaneous seizures in the Mongolian gerbil (*Meriones unguiculatus*). *Psychon. Sci.* 11:227–228.

Thiessen, D. D., G. Lindzey, and J. Nyby. 1970. The effects of olfactory deprivation and hormones on territorial marking in the male Mongolian gerbil (*Meriones unguiculatus*). *Horm. & Behav.* 1:315–326.

Thiessen, D. D., K. Owen, and G. Lindzey. 1971. Mechanisms of territorial marking in the male and female Mongolian gerbil (*Meriones unguiculatus*). *J. Comp. & Physiol. Psych.* 77(1):38–47.

Thiessen, D. D., F. E. Regnier, M. Rice, N. Isaacks, M. Goodwin, and N. Lawson. 1974. Identification of a ventral scent marking pheromone in the male Mongolian gerbil (*Meriones unguiculatus*). *Science* 184:83–85.

Thiessen, D. D., and M. Rice. 1976. Mammalian scent gland marking and social behavior. *Psych. Bull.*

Thiessen, D. D., P. Wallace, and P. Yahr. 1973. Comparative studies of the glandular scent marking in *Meriones tristrami*, an Israeli gerbil. *Horm. & Behav.* 4:143–147.

Thiessen, D. D., and P. Yahr. 1970. Central control of territorial marking in the Mongolian gerbil. *Physiol. & Behav.* 5:275–278.

Thiessen, D. D., P. Yahr, and G. Lindzey. 1971. Ventral and chin gland marking in the Mongolian gerbil (*Meriones unguiculatus*). *Forma et Functio* 4:171–175.

Thiessen, D. D., P. Yahr, and K. Owen. 1973. Regulatory mechanisms of territorial marking in the Mongolian gerbil. *J. Comp. & Physiol. Psych.* 82:382–393.

Thoday, A. J., and S. Shuster. 1970. The effects of hypophysectomy and testosterone on the activity of the sebaceous glands of castrated rats. *J. Endocrinol.* 47:219–224.

Thompson, J. A., and F. N. Pears. 1962. The functions of the anal glands of the brushtail possum. *Victoria Nat.* 78:306–308.

Thompson, R. W., and L. G. Lippman. 1972. Exploration and activity in the gerbil and rat. *J. Comp. & Physiol. Psych.* 80(3):439–448.

Thor, D. H. 1972. Schedules of self-lighting behavior: The yellow canary, goldfish and Mongolian gerbil. *J. Gen. Psych.* 87:23–35.

Tinbergen, N. 1957. The function of territory. *Bird Study* 4:14–27.

Tomkins, G. M., T. D. Gelehrter, D. Granner, D. Martin, Jr., H. H. Samuels, and E. B. Thompson. 1969. Control of specific gene expression in higher organisms. *Science* 166:1474–1480.

Topping, J. S., and J. M. Cole. 1969. A test of the odor hypothesis using Mongolian gerbils and a random trials procedure. *Psychon. Sci.* 17:183–184.

Udvardy, M. D. F. Mammalian evolution: Is it due to social subordination? *Science* 170:344.

Ungar, G., ed. 1970. *Molecular mechanisms in memory and learning.* New York: Plenum Press.

Vander Weele, D. A., and R. M. Abelson. 1973. Selected schedules of reinforcement in the Mongolian gerbil. *Psych. Reports* 33:99–104.

Vander Weele, D. A., R. M. Abelson, and J. Tellish. 1973. A comparison of ratio behavior in the gerbil and white rat. *Psychon. Soc. Bull.* 1(1B):62–65.

Vander Weele, D. A., and J. A. Tellish. 1971. Adipsic to polydipsic shift in gerbils induced by food deprivation. *Psych. Reports* 29:479–486.

Vick, L. H., and E. M. Banks. 1969. The estrous cycle and related behavior in the Mongolian gerbil (*Meriones unguiculatus*) Milne-Edwards. *Commun. Behav. Biol.* 3:117–124.

Vosseler, J. 1929. Beiträge zur Kenntnis der Fossa (*Cryptoprocta ferox Benn*) und ihrer Fortpflanzung. *Zool. Gart. NF.* 2:1–9.

Wales, N. A. M., and F. J. Ebling. 1971. The control of the apocrine glands of the rabbit by steroid hormones. *J. Endocrinol.* 51:763–770.

Wallace, P., K. Owen, and D. D. Thiessen. 1973. The control and function of maternal scent marking in the Mongolian gerbil (*Meriones unguiculatus*). *Physiol. & Behav.* 10:463–466.

Wallen, K., and S. E. Glickman. 1974. Effect of peripheral anosmia on ventral rubbing in the gerbil. *Behav. Biol.* 11: 569–572.

Walters, G. C., and E. L. Abel. 1971*a*. Effects of a marihuana homologue (Pyrahexyl) on avoidance learning in the gerbil. *J. Pharm. & Pharmacol.* 22: 310–312.

———. 1971*b*. Passive avoidance learning in rats, mice, gerbils and hamsters. *Psychon. Sci.* 22:269–270.

Walters, G. C., and R. D. Glazer. 1971. Punishment of instinctive behavior in the Mongolian gerbil. *J. Comp. & Physiol. Psych.* 75:331–340.

Walters, G. C., J. Pearl, and J. V. Rodgers. 1968. The gerbil as a subject in behavioral research. *Psych. Reports* 12: 313–318.

Walther, F. 1964. Verhaltensstudien an der Gattung Tragelaphus de Blainville, 1816, in Gefangenschaft, unter besonderer Berücksichtigung des Socialverhaltens. *Z. Tierpsychol.* 21:393–467, 642–646.

Webster, D. B. 1963. A sebaceous gland complex in the gerbil, *Meriones* (abstract). *Amer. Soc. Zool.* 3:488.

Webster, D. B., and K. L. Caccavale. 1966. Roles of eyes and vibrissae in maze running behavior of gerbils and kangaroo rats. *Amer. Zool.* 6:23.

Wechkin, S. 1964. Social dominance in the Mongolian gerbil (*Meriones unguiculatus*). *Ecol. Soc. Amer. Bull.* 50: 89–90.

Wechkin, S., and R. C. Cramer. 1971. The role of site familiarity in aggression toward strangers in the Mongolian gerbil. *Psychon. Sci.* 23:335–336.

Wechkin, S., and R. L. Reid. 1970. Social competition in the Mongolian gerbil (*Meriones unguiculatus*). *Psychon. Sci.* 19:285–286.

Werner, H. J., W. W. Dalquest, and J. H. Roberts. 1952. Histology of the scent gland of the peccaries. *Anat. Rec.* 113:71.

Whalen, R. E., and W. G. Luttge. 1971. Testosterone, androstenedione and dihydrotestosterone: Effects on mating behavior of male rats. *Horm. & Behav.* 2:117–125.

Whishaw, I. Q. 1972. Hippocampal electroencephalographic activity in the Mongolian gerbil during natural behaviours and wheel running and in the rat during wheel running and conditioned immobility. *Can. J. Psych./Rev. Can. Psych.* 26(3):219–239.

White, L. E., Jr. 1965. Olfactory bulb projections of the rat. *Anat. Rec.* 151:465–480.

Whitsett, J. M., and D. D. Thiessen. 1972. Sex differences in the control of scent marking behavior in the Mongolian gerbil (*Meriones unguiculatus*). *J. Comp. & Physiol. Psych.* 78:381–385.

Wilber, C. G., and R. D. Gilchrist. 1965. Organ weight: Body weight ratios in the Mongolian gerbil, *Meriones unguiculatus*. *Chesapeake Sci.* 6(2):109–114.

Williamson, A. R., and R. Schweet. 1965. Role of the genetic message in polyribosome function. *J. Mol. Biol.* 11: 358–372.

Wilson, E. O. 1966. Behavior of social insects. *Symp. Royal Entomol. Soc. Lond.* 3:81–96.

Wilz, K. J., and R. L. Bolton. 1971. Exploratory behavior in response to the spatial rearrangement of familiar stimuli. *Psychon. Sci.* 24(3):117–118.

Winkelmann, J. R. and L. L. Getz. 1962. Water balance in the Mongolian gerbil. *J. Mammal.* 43:150–154.

Wise, L. M., and E. Parker. 1968. Discriminative maze learning in the Mongolian gerbil. *Psych. Rec.* 18:201–203.

Wynne-Edwards, V. C. 1962. *Animal dispersion in relation to social behaviour*. New York: Hafner Pub. Co.

Yahr, P. 1972. Steroid regulation of territorial scent marking in the Mongolian gerbil (*Meriones unguiculatus*). Ph.D. dissertation, University of Texas at Austin.

————. 1976. Social subordination and scent marking in male Mongolian gerbils (*Meriones unguiculatus*). Unpublished.

————. In press *a*. The role of aromatization in androgen stimulation of gerbil scent marking. *Horm. & Behav.*

————. In press *b*. Effect of hormones and lactation on gerbils that seldom scent mark spontaneously. *Physiol. & Behav.*

Yahr, P., and S. Kessler. 1975. Suppression of reproduction in water-deprived Mongolian gerbils (*Meriones unguiculatus*). *Biol. Reprod.* 12:249–254.

Yahr, P., and D. D. Thiessen. 1972. Steroid regulation of territorial scent marking in the Mongolian gerbil. *Horm. & Behav.* 3(4):359–368.

————. 1975. Estrogen control of scent marking in female Mongolian gerbils (*Meriones unguiculatus*). *Behav. Biol.* 13:95–101.

Young, J. Z. 1950. *The life of vertebrates*. Oxford: Clarendon Press.

Zannier, R. 1965. Verhaltensuntersuchungen an der Zwergmanguste *Helogale undulata rufula* im Zoologischen Garten Frankfurt am Main. *Z. Tierpsychol.* 22:672–695.

Index

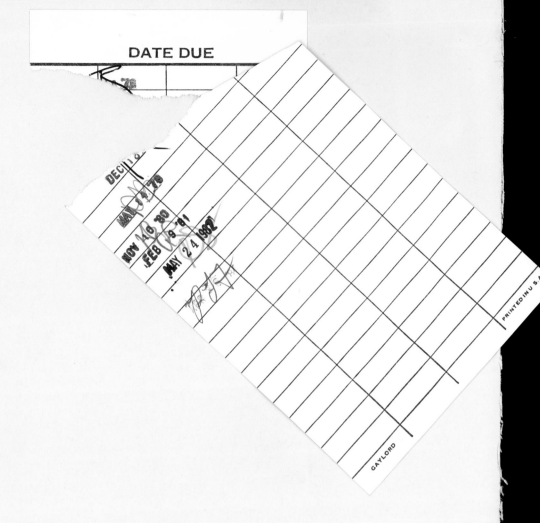